MODERN
GROUNDWATER
EXPLORATION

MODERN GROUNDWATER EXPLORATION

Discovering New Water Resources in Consolidated Rocks Using Innovative Hydrogeologic Concepts, Exploration, Drilling, Aquifer Testing, and Management Methods

Robert A. Bisson
Earthwater Technology International, Inc.

Jay H. Lehr
The Heartland Institute

WILEY-INTERSCIENCE

A JOHN WILEY & SONS, INC., PUBLICATION

For general information on our other products and services please contact our Customer Care Department within the U.S. at 877-762-2974, outside the U.S. at 317-572-3993 or fax 317-572-4002.

Wiley also publishes its books in a variety of electronic formats. Some content that appears in print, however, may not be available in electronic format.

Library of Congress Cataloging-in-Publication Data:
Bisson, Robert A.
   Modern groundwater exploration : discovering new water resources in consolidated rocks using innovative hydrogeologic concepts, exploration, drilling, aquifer testing, and management methods / Robert
A. Bisson, Jay H. Lehr.
      p. cm.
   ISBN 0-471-06460-2 (cloth : alk. paper)
   1. Groundwater—Research—Methodology. 2. Prospecting—Geophysical methods.
I. Lehr, Jay H., 1936– II. Title.
GB1001.7 .B57 2004
628.1′14—dc22

2003023862

Printed in the United States of America

10 9 8 7 6 5 4 3 2 1

*"... The ocean is a desert with its life underground
and a perfect disguise above ..."*

from "A Horse With No Name," Dewey Bunnell, 1971

Contents

Preface

This book is about the continued relevance of exploration and discovery in improving the water shortages faced by a thirsty world. While the popular press seems mesmerized by ethical debates concerning quantum leaps in our understanding of the human genome and the worldwide blame-game over global warming, equally significant discoveries in the energy and groundwater fields resulting from modern geophysical exploration concepts and technologies have been largely ignored. This is a story of the vital importance of iconoclasm in modern science—a story of how a few people thinking outside the box can make important contributions to the world.

No one knows how many unexplored and unexploited aquifers exist, but the amount of water stored in them is thought to be considerable. The focus of this book is to help exploration geologists around the globe to uncover vast stores of water yet undiscovered. Water exploration must tie itself to the many great advances made by the petroleum industry, which long ago tied itself to computer technology. Today we stand on the launching pad to a journey into high technology groundwater location and development whose foundation has been laid for us through advances in petroleum and other mineral explorations.

In a brilliant article by Jonathan Rauch in the January 2001 issue of *Atlantic Monthly* ("The New Old Economy: Oil, Computers and the Reinvention of the Earth"), the author paints a most optimistic picture of future oil availability. He predicts "that the demand for oil will peter out well before any serious crimp is felt in the supply" because "something cheaper and cleaner . . . will come along, and the oil age will end with large amounts of oil left unwanted in the ground."

Rauch arrives at this belief because he likens the energy industry to the computer industry. He says, "Most people understand intuitively that the essential resource in Silicon Valley is not magnetic particles on floppy discs or hard drives in servers, or lines of code or bits of data; it is human ingenuity." He concludes, ingeniously, that "knowledge, not petroleum, is becoming the critical resource in the oil business; and though the supply of oil is fixed, the supply of knowledge is boundless."

In every sense except the one that is most literal and least important, the planet's resource base is growing larger, not smaller. Every day the planet becomes less an object and more an idea.

It is difficult to fully understand why groundwater stands alone as a resource whose cost has risen rather than fallen in the crosshairs of man's ingenuity. It may be that water has almost always been controlled by governments rather than private industry, thereby reducing creative solutions. It may be that too little research funding was afforded groundwater because, although water's intrinsic value was recognized, its true economic value was, and largely still is, maintained artificially low through government subsidies. It is also true that the international hydrology community impeded innovation as much as conformist geologists resisted plate tectonics only half a century ago.

Whatever the reason, we sincerely hope that the information, the concepts, the data, and the achievements we present on the following pages break up the logjam of progress that has held back the optimization of groundwater development the world over.

Robert A. Bisson
Earthwater Technology International, Inc.
www.watermap.com
Jay H. Lehr
The Heartland Institute
www.e3power.com

Acknowledgments

First and foremost, the authors offer our sincere gratitude to Sabine Bisson and Janet Lehr for their steadfast support and forbearance during the production of this book.

We are most appreciative of the chapter contributions of Mr. Alexander Raymond Love (Chapter 1), Dr. Utam Maharaj (Chapter 4), and Dr. Ewan Anderson (Chapter 6), and also recognize co-authors of primary source materials summarized in Chapters 3, 4, and 6 of the book, including Dr. Roland B. Hoag (original technical reports on Somalia, Sudan, and Trinidad), Joseph Ingari (original technical reports on Somalia and Sudan), Dr. Farouk El-Baz, David Hoisington, Dennis Albaugh (original technical reports on Somalia and Sudan), and Dr. Mohamed Ayed and Dr. Farouq Ahmed (original technical reports on Sudan). Special thanks are extended to James Jacobs for his technical review and editing of the book manuscript and to Richard Cooper for his on-target Dewey Bunnell quotation.

We applaud the visionary leadership of Warren and Lisa Wiggins of the New Transcentury Foundation (NTF) and former USAID East and Southern Africa REDSO Director and Sudan Mission Director John Koehring in daring to apply new technologies to solving Africa's chronic water-related catastrophes. The authors offer special recognition to Trinidad water well-drilling entrepreneur Lennox Persad, who risked millions of his own money to help ensure the security of his country's future water supply. We also acknowledge former chairman Nazir Khan, former CEO Kansham Khanhai, Water Resources Agency Director Dr. Utam Maharaj, and Tobago Regional Manager Oswyn Edmund of the Water and Sewerage Authority of Trinidad and Tobago (WASA), for their leadership in opening the way for breakthrough water discoveries employing modern groundwater exploration methods, along with current CEO Errol Grimes for his dedication to exporting WASA's new knowledge base to other Caribbean states. The megawatersheds projects implemented by NTF, USAID, and WASA represent the bookends of major groundwater discoveries over the past twenty years.

Robert Bisson wishes to thank his many other mentors, teachers, partners, employees, and clients, who collectively made possible the advancements of techniques and development of new concepts that led to the fresh body of knowledge reflected in the case study summaries of this book, including Col. Reginald Bisson, Adrienne Bisson, Dr. Roland B. Hoag, Jr., Joseph C. Ingari, Peter D. Hofman, Gary Armstrong, Larry Whitesell, Hon. John H. Sununu, William Goldstein, Dr. Hasan Qashu, Captain Jacques-Yves Cousteau, Dr. William K. Widger, Jr., Dr. Olaf R. McLetchie, Dr. Lincoln R. Page, Robert M. Snyder, Alan Vorhees, Donald Gevirtz, Robert Kaufman, Kermit Roosevelt, Jon Armstrong, Dr. Tony Barringer, Robert Porter, Dr. Compton Chase-Lansdale, Donald Jutton, Frank Kenefick, Hon. Kevin Harrington, Judge J. John Fox, James Lee, Douglas Lee, Arthur Lipper, Ian Sims, and Dr. Dennis Long.

Jay Lehr wishes to thank posthumously Dr. Joseph Upson, his first boss and mentor at the U.S. Geological Survey in 1955. When asked "what are the working hours" by this young scientist, he responded by saying, "We start promptly at 8:30 AM and we leave when the job is done." Happily, because of Joe I learned that the job is never done and, as this book reflects, the job in groundwater science is still ongoing.

Introduction

It was only in the last quarter of the twentieth century that the subject of groundwater was transformed from its nineteenth century place as a popular topic of mystical origins to a key economic commodity and political issue scrutinized by modern exploration scientists. As a consequence, the world in general knows little more now about the nature and extent of its fresh groundwater resources than it did at the turn of the nineteenth century. At the same time, governments and the scientific community do today agree on one point—that groundwater can no longer be considered a resource to be freely and blindly extracted and subjected to contamination with impunity, but rather must be "managed" and "protected" as an endangered resource.

The fatal flaw in this well-intentioned action plan is lack of data. Even basic water-related data, from precipitation to runoff, are sparse and often inaccurate, and far less is known about the water-related properties of underground environments. This holds true as much in developed countries as in the third world. "If you can't measure water, you can't manage it," observed Arthur Askew, director of hydrology at the World Meteorological Organization (WMO) in the August 9, 2002, issue of *Science*. Unfortunately, while ill-planned and poorly documented water well-drilling schemes have proliferated worldwide, governments and international development institutions have not historically supported scientifically designed hydrogeological surveys and groundwater data logging, and private sector profit incentives to perform those tasks have been sorely lacking. As a consequence, very little data-driven published or archival material exists about measured quantities of global groundwater resources, and critical water-related decisions are being made behind a curtain of ignorance that adversely affects the global economy and the lives of all of us.

The coincidence of space-age technological innovations and scientific advancements has opened a window of opportunity to bring the groundwater knowledge base up to par with other strategic minerals in the first quarter of the twenty-first century. Governments and financial institutions are turning fresh water into a market-value commodity by privatizing water utilities. This being the case, fresh water stored and flowing beneath the ground should rightly be classified as an "economic mineral" and be thoroughly investigated by economic geologists using modern geologic concepts and all of the space-age exploration tools that revolutionized the minerals industry over the past forty years.

This book presents readers with solid data supporting a new paradigm that vastly increases estimates of groundwater balance along with a description of the modern exploration and development methods required to measure and access these untapped resources.

1 A Historical Perspective

In humid regions, primitive humans paid little attention to water. It was always present and, like air, was taken as a matter of course. However, in semiarid and arid regions, the occurrence of water controlled the activities of humans. Villages were originally built on perennial streams or around water holes. Our early movements consisted chiefly of migrations to perennial water in the dry season and ventures into new pastures or hunting grounds in the wet season.

Primitive humans learned to dig for water, possibly by observing the actions of wild horses and wolves in search of water. As soon as we learned to domesticate and rear cattle and sheep, the water well became the most important possession.

The Bible described many incidents illustrating the importance of groundwater supplies to the tribes of Israel. Abraham and Isaac were renowned for their success at constructing wells. The Father of Modern Hydrology, O.E. Meinzer once said that the twenty-sixth chapter of Genesis read like a water-supply paper. Most people recall from the Old Testament how the Jews suffered for want of water in their 40 years of wandering in the deserts. To quell a near revolt by his people, Moses smote a rock with his rod and a fountain of water burst forth.

The ancient Greeks in the early seventh century BC told the story of Tantalus, Zeus' favorite mortal son who stole the ambrosia and nectar from the gods that gave the gods endless lives. Tantalus tried to share the heavenly food with mortals to give humans immortality. Zeus punished Tantalus by hurling him to Tartarus, a prison of darkness where Tantalus currently stands, trapped in the pool of water that is chin-height. He cannot drink it, though, for anytime he lowers his mouth to take a drink, the water recedes. The ancient Greeks knew the value of water, for Tantalus was sentenced to an eternal life of thirst, the most terrible punishment available. Hence, the word, tantalize.

Modern Groundwater Exploration: Discovering New Water Resources in Consolidated Rocks Using Innovative Hydrogeologic Concepts, Exploration, Drilling, Aquifer Testing, and Management Methods, by Robert A. Bisson and Jay H. Lehr
ISBN 0-471-06460-2 Copyright © 2004 John Wiley & Sons, Inc.

The Romans depended on many shallow wells and springs before they built their first aqueduct in 312 BC. The soil was so rich in springs and underground streams that wells could be sunk successfully at any point, and the average depth necessary was only about 5 m. Such wells were common from the earliest period, of the Roman Empire Excavations in the Roman forum have uncovered more than 30 wells dating back to the Republic.

The drilling rather than digging of artesian wells in France and Italy began in the twelfth century and created considerable popular and scientific interest on the occurrence of underground water. The art of drilling and casing wells was actually invented, perfected, and extensively practiced by the ancient Chinese. They used bamboo poles and patience to penetrate hundreds of feet. Wells were started by the grandfather and completed by the grandson.

The most extraordinary works of ancient humans for collecting groundwater are the qanats and karezes of the Persians and Afghanies. The qanats and karezes are tunnels that connect the bottoms of shafts, which were dug by humans working as moles over long periods of time and are conspicuous over all the high central valleys of Iran. Thirty six of these tunnels supplied Teheran and the highly cultivated tributary agricultural area.

In ancient times, springs were considered the miraculous gifts of the gods; they wrought miracles and consequently were places where temples were built. These superstitions continue today with those who optimistically overestimate the therapeutic value of medicinal springs.

Prior to the latter part of the seventeenth century, it was generally assumed that the water discharged by the springs could not be derived from the rain, first because the rainfall was believed to be inadequate in quantity and second, because the Earth was believed to be too impervious to permit penetration of the rain water far below the surface. With these two erroneous postulates lightly assumed, the philosophers devoted their thought to devising ingenuous hypotheses to account in some other way for the spring and stream water.

Two main hypotheses were developed: one to the effect that sea water is conducted through subterranean channels below the mountains and is then purified and raised to the springs and the other to the effect that in the cold dark cavern under the mountains, the subterranean atmosphere and perhaps the Earth itself are condensed into the moisture. The sea water hypothesis gave rise to subsidiary ideas to explain how the sea water is freed from its salt and how it is elevated to the altitude of the springs. The removal of the salt was ascribed to processes of either naturally occurring distillation or filtration.

Beginning with the middle of the sixteenth until the close of the seventeenth century, numerous publications appeared that contained discussions of groundwater, but the two ancient or classic Greek hypotheses chiefly occupied the field, although an infiltration theory was explained in 1580 by Bernard Palissy. In the later part of the seventeenth century, Perrault, Mariotte, and Halley abandoned the theories of the past and actively undertook experimental work to determine the source and movements of groundwater, and thus was born the science of groundwater. Perrault made rainfall measurements during three years and roughly estimated the area of the drainage basin of the Seine River above a point in Burgundy and of the runoff from this same basin. He computed that the quantity of water that fell on the basin as rain or snow was about six times the quantity discharged by the river. Crude as his work was, he definitely demonstrated the fallacy of the old assumption of the inadequacy of the rainfall to account for the discharge of springs and streams.

Mariotte computed the discharge of the Seine at Paris by measuring its width, depth, and velocity at approximately its mean stage and by doing so verified Perrault's results. About the same time, Halley made crude tests of evaporation and demonstrated that the evaporation from the sea is sufficient to account for all the water supplied to the springs and streams, thus removing the need for any other mysterious subterranean channel to conduct the water from the ocean to the springs.

Centuries were required to free scientists from superstition and wild theories handed down from earlier generations regarding the unseen subsurface water. To a certain extent, we still live at a time when great misunderstanding if not superstition exist with regard to the occurrence and movement of groundwater. The elementary principle that gravity controls motions of water underground as well as at the surface is still not appreciated by all engaged in the development of the world's vast groundwater supplies.

Many people still believe that the magical forked witch stick is able to point to underground water streams and will actually twist in the hands of the operator in its endeavor to do so.

These popular superstitions are examples of the ability to believe without the foundation of facts, and this peculiar ability exists in the minds of both educated and uneducated men and women. Inasmuch as the movements of underground water cannot be observed at the surface, they have been subject to wild speculation. Even an American judge in a court case once ruled that "percolating water moves in a mysterious manner in courses unknown and unknowable."

Little by little in the last decades of the twentieth century, groundwater hydrologists dragged the water supply fraternity and the public at large kicking and screaming into a twenty-first century. Now groundwater resources are appropriately valued as often the best hope for enabling society and commerce to move forward unhindered by water shortages. Forty-seven percent of the U.S. population now depends on groundwater for its drinking water. In the Asia-Pacific region, 32% of the population is groundwater dependent; in Europe, 75%; in Latin America, 29% and in Australia, 15%.

The authors of this book played their role of ardent enthusiastic scientists during this period battling ever-present opposition to belief that significant quantities of groundwater supply could be sustained. Although we approached success in our efforts, it was still a small victory as our intent has been to reveal to the world the vast quantities of groundwater yet hidden deep within the Earth, often beneath arid lands. Thus far, there has been little confidence in our conceptual model or paradigm.

We have long believed that a planet whose surface is covered by water should not be facing water shortages. Admittedly, 97% of the Earth's water is too salty for humans and agriculture, and glaciers and ice caps put another significant portion out of reach. But we have long believed that a significant portion thought to be out of reach under the ground is not.

Energy-intensive desalting of seawater is currently too expensive except in wealthy but dry areas near seacoasts. Our fresh surface water has been allocated in most of the developed world, with Canada being a rare exception.

Although humans have learned well over the past century to conserve water in such that water use per person has actually declined, the addition of the final two billion people on the planet in the next 40 years before its population stabilizes (in accordance with most sound demographic projections) will require considerable additional water supplies. If we fail to develop additional water supplies, international strife will remain. Half of our continental land lies within river basins shared by more than one country. Multinational water claims have not and likely will not provoke war, but local and regional conflicts have occurred over inequitable allocation and use of water resources. International diplomacy commonly encourages opposing countries to cooperate, but not always before lives are lost. Most recently, apartheid battles in South Africa in 1990, Iranian and Iraqi disputes in 1991, and intrastate conflicts in India in the mid 1990s cost thousands of lives.

There has been an explosive development of groundwater in several major deserts of the world in the last half of the twentieth century. Preliminary results of activity in the Sahara and throughout the Arabian Peninsula substantiate the occurrence of vast amounts of water stored beneath desert lands. This development is due to efforts of groundwater geologists and engineers, well construction crews, and political leaders who had the courage to launch the investigations against accepted water resources paradigms.

Several factors make water development programs in arid regions feasible:

(1) The deserts offer uncrowded space
(2) Favorable climate for nearly year-round crop growth
(3) Large areas of reasonably good soils and food-fiber requirements for persons in mineral and petroleum resource industries in desert areas.

Throughout most of the world, aquifers have not been regarded as true water resource reservoirs. Rather, they are simply viewed as holding tanks for annual contributions of what is unfortunately thought of as safe yield. The annual increment of groundwater is skimmed off the top when the basin below is depleted in any significant way. But in fact, the surface-water reservoir that remains full is obviously as poorly managed as that which remains empty; so too is the groundwater reservoir poorly managed when it is not allowed to rise and fall in contrast to the vagaries of the natural cycle and the demands of the human population.

Good water management is the optimum manipulation of the available water resource to serve the greatest common good. It includes the coordination of both the natural aspects of the hydraulic cycle and every artificial operation that can be performed upon it, save, in most cases, the drastically expensive and uneconomic interbasin transfer of surface water.

The concept of the hydrologic cycle has become so generally accepted that it is difficult to appreciate the long history that lies back of its development and demonstration, from the dawn of history until comparatively recent times, barely a quarter of a century ago. The central concept in the science of hydrology is the hydrologic cycle, a convenient term to denote the circulation of the water from the sea, through the atmosphere, to the land and thence with numerous delays back to the sea by overland and underground routes and in part through re-evaporation and transpiration from vegetation, lakes, and streams. It involves the measurement of the quantities and rates of movement of water at all times and at every stage of its course through multiple reservoirs, from a height of 15 km above the ground to a depth of some 5 km beneath it. The reservoirs include atmospheric moisture, oceans, rivers, lakes, icecaps, soil, and groundwater. The transport mechanism from one physical state or aquifer to another is either gravity or solar energy over periods that range from hours to thousands of years.

The pioneer of intensive groundwater investigations was Germany's Adolph Theim who introduced field methods for making tests of the flow of groundwater and applied the laws of flow in developing water supplies. Under his influence, Germany became the leading country in supplying its cities with groundwater, and it still derives over 80% of its needs from wells.

Because we can see surface waters and because such tremendous amounts of money have been spent in building visible dams, levees, artificial reservoirs, aqueducts, and irrigation canals involving surface water, it is openly natural that we tend to think of that water as the major source of the world's needs. Actually less than 3% of unfrozen fresh water available at any given moment on our planet Earth occurs in streams and lakes. The other more than 97%, estimated at eight trillion acre-feet, is underground.

The total amount of water on our planet has almost certainly not changed since geological times. Water can be polluted, abused, and misused, but it is neither created nor destroyed; it only migrates.

Groundwater is tracked by remote sensing and tracer techniques, but the water movement is exceedingly difficult to follow. It is known that groundwater migrates slowly. Sometimes groundwater moves only a few millimeters a day, although occasionally it is a few meters per day. Near the water table, the average cycling time of water may be a year or less, whereas in deep aquifers, it may be as long as thousands of years. It is easier to measure water tables. Through test wells and controlled pumping, it is not difficult to measure the recharge rate and flow behavior around a particular site. The difficulty comes in sensing movement in the aquifer as a whole. Water can be stored in the pores of rocks, but it can also be stored in cracks and fractures, and sometimes these fissures can provide conduits to allow water to travel quickly and over great distances.

No one knows how many unexplored and unexploited aquifers exist, but the amount of water stored in them is thought to be considerable. This is the focus of this book, to help exploration geologists around the globe to uncover vast stores of water yet undiscovered. Water exploration companies have not yet made a major impact on the stock markets of developed countries, but they would not be a bad bet for adventurous investors.

Water exploration must tie itself to the many great advances made by the petroleum industry, which long ago tied itself to computer technology. The petroleum industries need for processing power is insatiable, and it has resulted in many computer technology advances. Today we stand on the launching pad to a journey into high technology groundwater location and development whose foundation has been laid for us through advances in petroleum and other mineral explorations. The oil industry itself has been a driving force in the computer industry where Texas Instruments began as a company in 1930 known as Geophysical Service. Seismic imaging now available for groundwater studies led to the development of computer programs to assess sound waves generated in rock to infer the nature and location of rock layers capable of trapping oil. From initial two-dimensional images, computers ultimately were taught to process gigabytes of data that would result in three-dimensional images.

In 1985, more than a day of computing time was required to analyze a square kilometer of subsurface structure; by 1995, computers could do it in 10 minutes, and the cost to survey 10km^2 dropped from millions of dollars to tens of thousands of dollars.

Concurrently, computer-assisted drilling technology advances include saw drill bits that have direct sensing tools to evaluate physical channels, and electrical characteristics of what they were drilling through while transmitting their exact location to the surface and enabling the implementation of immediate course corrections. In some ways, it is truly amazing that this book is only being written at the beginning of the twenty-first century. Advances to be described in the following chapters regarding groundwater development were recognized in the parallel fields of geologic science and engineering, more than two decades ago. This is why, with the exception of water every mineral resource on Earth has become less and less expensive than it was in our youth. Technologic- and knowledge-based advances have reduced the costs of location, development, and refinement of every other mineral in the Earth, without exception.

Mines and oil fields once abandoned have been reopened for redevelopment of formerly uneconomic resources. Groundwater, however, has been saddled with a century old paradigm. We place a straw in only the upper portion of our water-filled glass (or aquifer) ignoring its deeper regions because of our inability to discern deeper rock structures. Or we are misguided by beliefs that recharge could not continuously replenish the deeper portions of that or any aquifer. So leave it there lest we become dependent on a nonrenewable resource. A similar philosophy would have left us in the Stone Age, never to develop an Iron Age or use any other minerals in the Earth's crust.

At the same time, commitments are being made by the United States and other governments that could severely damage independent efforts to discover the realities of deep groundwater environments. For example, in recent years, the U.S. Environmental Protection Agency and similar agencies worldwide not only missed an ideal opportunity to advance the understanding of deep groundwater resources, but delayed its advancement by demanding instant competence of an unprepared scientific community. In addition, U.S. EPA spent billions of dollars on groundwater "cleanup" and "protection" without first performing the due diligence required to critically evaluate the actual knowledge base. The origins of this unfortunate situation are of recent vintage. In the 1980s, it was disconcerting for exploration scientists to observe the U.S. Congress balk at a request from the world's preeminent institution of basic geological research and knowledge, the U.S. Geological Survey, for modest funding to update decades-old, low-resolution, pre-space-age geological maps (upon which groundwater studies are usually based); and it was even more bewildering in the 1990s to witness Congress funding a multi-billion-dollar nationwide groundwater cleanup effort without benefit of the requisite knowledge base they failed to develop a decade before.

Spending those billions of dollars have not only failed to advance the knowledge base about deep groundwater, but also have induced a surfeit of numerical models largely based on anachronistic concepts of groundwater occurrence. Such elegant, but ill-conceived models have been combined with sophisticated computer visualization programs and published as factual representations of global groundwater occurrence in professional journals and the popular press, leading engineers, economists, and political leaders to premature and erroneous conclusions regarding groundwater balance and the Earth's fresh water balance as well.

2 Megawatersheds— A New Paradigm

ALEXANDER RAYMOND LOVE

Founding Executive Director, Partnership to Cut Hunger and Poverty in Africa

"The originator of a new concept . . . finds, as a rule, that it is much more difficult to find out why other people do not understand him than it was to discover the new truths."

—Herman von Helmholtz

In 1875, a 17-year-old German student had just graduated from Gymnasium and was about to enter University. Intrigued with his early studies of physics, he had decided to pursue it as a career. He thus approached the head of the physics department at the university for advice. The professor was not encouraging. "Physics is a branch of knowledge that is just about complete. The important discoveries, all of them, have been made. It is hardly worth entering physics anymore."

The attitude of the professor was probably representative of the prevailing wisdom in "Newtonian or Classical" physics in 1875. It represented a plateau of knowledge that had been reached in the nearly two centuries since the early discoveries of Sir Isaac Newton. Little did the professor realize that the field of physics was on the verge of a major new age of discovery and development, one that would reshape the study of physics and profoundly impact on world development.

The professor, of course, also had no way of knowing that the student, Max Planck, would ignore his advice and go on to become a world famous physicist. Planck would introduce the quantum theory and help usher in the new age of physics with the publication of his

Modern Groundwater Exploration: Discovering New Water Resources in Consolidated Rocks Using Innovative Hydrogeologic Concepts, Exploration, Drilling, Aquifer Testing, and Management Methods, by Robert A. Bisson and Jay H. Lehr
ISBN 0-471-06460-2   Copyright © 2004 John Wiley & Sons, Inc.

work on black body radiation at the beginning of the twentieth century. Albert Einstein, then 21 years old, would eventually build on Planck's work and further refine the quantum theory, eventually to supplement it with his new theory of relativity.

The quote by Herman von Helmholtz and the story of young Max Planck are both from Barbara Lovett Cline's 1965 book *Men Who Made A New Physics*. I have read the book many times and continue to marvel at the process of rapid technological change that took place over such a relatively short period of history. It was indeed a true paradigm shift. To a layman, it seems not so much that Newtonian physics was wrong, as it was incomplete. Physicists continued to find unexplained phenomenon and inconsistencies with classical physics. Pursuit of these questions eventually opened new fields of opportunity in physics.

There are two key themes embodied in these excerpts from Cline's book that are pertinent to the telling of the story on megawatersheds. The first is reflected in the story of Max Planck. Here we see a leading expert in a field of technology, the professor of physics, defending the status quo base of knowledge and rejecting the hypothesis that new discoveries might usher in a new era. In this book, the reader will encounter numerous examples of this principle. In many cases, this inherent resistance to change has led to erroneous assessments of water availability and flawed investment decisions or recommendations. This was graphically illustrated in the cases of the Trinidad and Tobago water development programs discussed in Chapters 5 and 6.

More importantly, however, the general failure of the world at large to take advantage of the potential of the new megawatershed technology has curtailed the development of additional available water resources. This failure can have major economic, political, and developmental implications for future development of our world's water resources.

The second principle is closely related and is reflected in von Helmholtz's quote. It reflects his apparent frustration with the difficulty of convincing his colleagues of the validity of his new discoveries. Such is the case with the megawatershed paradigm. Skepticism remains strong among members of the technical community. This contributes to hesitation associated with risk aversion among technical experts, investors, and public bureaucrats alike. This book attempts to address these concerns in two ways. First, the telling of the successful case studies such as Sudan, Trinidad and Tobago, Somalia, and so on, serve to help prove the principles of megawatersheds by real-world example. Secondly the book discusses and illustrates how risk-sharing models can be used to shift the risk to technically knowledgeable entrepreneurial investors who operate in a different risk/benefit framework than the public sector. Both approaches must be followed if scepticism is to be overcome and a more rapid application of the megawatershed concept is to be achieved.

GLOBAL IMPLICATIONS

The premise of this book is that the megawatershed model can open up access to substantial quantities of water either not known to exist or thought to be fossil, nonrenewable, or inaccessible. In the introduction to this book, Robert Bisson indicates from results of several case studies that "ten times more fresh water than previously calculated may be in active groundwater circulation under drought plagued cities and villages across America, Asia, and Africa." To put this contention in perspective, we will review the global distribution of water availability. Most, of course, is in the oceans: 1.33 trillion km^3 or 96.5%. With an additional nearly 1% in saline/brackish groundwater or salt-water lakes [Shiklomanov, 1993]. Only 2.5% or 35 million km^3 is fresh water, and 70% of this is tied up in glaciers and permanent snow cover. Fresh groundwater accounts for 10.5 million km^3 or the bulk of the balance. This far exceeds the 0.091 million km^3 in freshwater lakes and the 2.1 million km^3 in rivers.

Given the relative magnitude of these figures, it seems apparent that any technological advance that offers greater access to the predominant source of fresh water (groundwater) should be welcomed with open arms. But the real significance probably lies in local application of the megawatershed principles in those locations where the economic and environmental costs of traditional technology and documented water availability fail to meet the needs of the local population. The case examples in this book highlight a number of such instances.

There are a number of overlapping themes woven throughout these case studies that can be reviewed as a prelude to the review of the cases themselves. They are as follows:

Megawatersheds and Economic Development: The Millennium Development Goals

Megawatersheds and Food Production

Megawatersheds and Environmental Needs

Megawatersheds and the Water Needs of Refugees and Drought Victims

Megawatersheds and Commercial/Industrial Use

Megawatersheds and International Political Challenges
Megawatersheds, Skepticism, and Risk Reduction

Finally, it is the review on increasing the application of the megawatershed concept to address some of the world's looming water shortages.

Megawatersheds and Economic Development: The Millennium Development Goals

The concept of megawatersheds, of course, applies to all of the inhabited continents of the Earth and many small island states, developed and underdeveloped (SIDS). However, the major unmet needs for water supply are in the developing world. It is there that both the quantity and quality of water supply are most inadequate and the cost of large surface water development schemes is prohibitive. A good place to start the analysis of water's key role in development is with the existing global compact for development reached at the United Nations. The UN millennium Declaration, "Development and Poverty Eradication," approved in 2000, set forth 8 development goals and 18 specific targets to be achieved by 2015. A number of these are directly related to adequate water supply. Access to clean water and sanitation is at the top of the list. But water is also key to the goals of reducing poverty and hunger by 2015 because of its critical role in agriculture and agriculture's key role in poverty reduction. Water is also critical to achieving overall goals in the environmental arena. Related goals in the health area such as reducing infant and child mortality are further dependent on water supply.

Drought and famine were not specifically targeted in the Millennium compact. However, drought and famine will require new approaches for providing water to drought victims and refugees. Ewan Anderson points this out elsewhere in this book. The UN and the development community have been negligent in not directly including elimination of manmade and natural disasters as key objectives in the overall program to reduce poverty. Such exclusion belies the reality of the development challenge in areas like Afghanistan and Sub-Saharan Africa today.

The UN has also declared access to water to be both a "right" and a "basic need": "The human right to water is indispensable for leading a life in human dignity. It is a prerequisite for the realization of other human rights."

It is clear from the UN documentation and related studies that water is critical to achieving many of the agreed economic and social objectives in the Developing World. Yet, as the World Panel on Financing Water Infrastructure points out "water has been underemphasized and neglected in the past, compared to other sectors." "The costs of neglect . . . are cumulative," as the report rightly points out. The panel's report "Financing Water for All" provides a comprehensive look at the challenges to financing water projects. It does not attempt to address new technological approaches to overcoming drinking water shortages. It does, however, provide a good overview of who owns, who manages, and who finances water systems. This perspective is of great importance to the discussion of megawatershed technology and the associated concept of risk reduction discussed later.

In this framework, the introduction of a new technical approach, e.g., the megawatershed concept, can have a far-reaching impact on a developing country's progress in reducing poverty, improving health, preserving the environment, and contributing to new approaches for dealing with drought and refugees.

A compelling argument exists today for moving megawatershed's exploration up on the priority list on the world's development agenda. In the following sections, we will touch on some of these specific development challenges in more detail.

Megawatersheds and Food Production

In 2002, the International Food Policy Institute and the International Water Management Institute published a comprehensive volume entitled *World Water and Food to 2025*, which analyzes the challenges to international food security posed by competition for water sources and water scarcity. The first chapter sets the stage for the premise of the book by stating, "The story of food security in the 21st century is likely to be closely linked to the story of water security. In coming decades, the world's farmers will need to produce enough food to feed many millions more people, *yet there are virtually no untapped, cost-effective sources of water for them to draw on as they face this challenge*. Moreover, farmers will face heavy competition for this water from households, industries, and environmentalists." (Emphasis added.)

The book highlights the fact that nearly 250 million hectares of land are under irrigation today, five times the amount under irrigation at the beginning of the twentieth century. Moreover, much of the successful increases in food production were associated with the technological breakthroughs in plant technology that came to be known as the Green Revolution. It is well to remember that the achievements of the Green

Revolution were closely associated with the increase in irrigation as well as the increased use of fertilizers. But the Green Revolution has had much less success in the arid and semi-arid areas of Asia and Africa where many of the world's poorest and most malnourished people or citizens reside. Irrigated agriculture uses 80% of global water supply and 86% of developing country water [IFPRI].

J.A. Allen in his book, *The Middle East Water Question/Hydropolitics and the Global Economy*, highlights this heavy demand for food production by comparing the components of individual water use per year:

Drinking: $1\,m^3$ per year

Other domestic uses: 50 to $100\,m^3$ per year

Embedded in food: $1000\,m^3$ per year, either from naturally occurring soil water or from irrigation systems.

The pursuit of greater food production will thus be closely associated with the competition for water. Already it is clear that major new infrastructure projects, such as storage dams, will be subject to greater scrutiny and in some cases blocked from implementation, as happened with the proposed dam on the White Nile in 2003. Moreover, existing irrigation schemes are subject to waterlogging and salinity problems that require costly new drainage schemes to maintain production levels. Rapid urbanization is adding to the competition with agriculture for water. The environmental community is actively competing for water to support "environmental" needs, including the maintenance of wetlands, natural forest reserves, aquatic species, and so on.

What does the megawatershed paradigm offer to meet this challenge? First, there is a need to better understand the character of groundwater resources in areas already irrigated with groundwater. In many cases, megawatershed principles could help better understand the source, character, and sustainability of existing systems and allow more effective long-term use. Secondly, there are major known aquifers (e.g., North Africa's Nubian Sandstones) that are underutilized because of the prevailing "fossil" groundwater model that does not recognize active recharge to these vast groundwater systems. A better understanding of these aquifers might open up new resources that would help minimize competition with agriculture. Third, much of the problem in food security centers on the populations in arid and semi-arid areas. The inability to generate income from erratic agriculture as well as food shortage is at the root of the problem. Supplemental water from new strategically developed megawatersheds could help meet food needs, especially during periods of drought when crops fail. This is closely related to drought/refugee issues. It is also closely related to the complex factors that motivate urban migration, e.g., the lack of adequate income opportunities in rural areas where agriculture and related off-farm employment have declined.

Afghanistan presents a promising case study on the application of megawatershed theory to increase agriculture development. Traditionally, the Afghans have tapped surface water and shallow aquifers to supplement rain-fed agriculture. The use of traditional kerezes attests to the commitment and ingenuity of the Afghans in searching for additional agriculture water. The kerezes are ingenious tunnel systems extending into the mountains to tap groundwater sources, some of which are reportedly 20 miles long.

Today, amidst the devastation of decades of war, restoring agriculture has become a top priority; water supply from traditional groundwater sources is proving inadequate, and new sources must be found. The heavily faulted, fractured, and mountainous physiography and high-elevation snow pack and precipitation of Afghanistan provides conditions conducive to locating substantial additional groundwater through the application of megawatershed technology.

Discussions are currently underway regarding the initiation of a countrywide megawatershed assessment of Afghanistan to locate large sources of additional groundwater.

Before we leave the subject of food production, let us take another look at the phenomenon called the "Green Revolution." The phrase Green Revolution refers to the breakthrough in agriculture that took place in the 1960s. At that time, the major agriculture research stations at Los Banos in the Philippines, the International Rice Research Institute (IRRI), and the International Maize and Wheat Improvement Center (CIMMYT) in Mexico developed new high-yielding varieties of rice, wheat, and maize. These three cereal crops are the basic food staples for a large percentage of the world's population. This is especially true in the developing world. Before the technological breakthrough, there was a growing consensus that the world's burgeoning population would outstrip food production in a Few decades and growing famine would follow.

The development of new cereal varieties associated with improvements in fertilizer use, pesticides, and water control ushered in a true paradigm shift in the world's food production. Today, nearly 40 years later, the world still produces enough food for a substantially larger population. Famines in South Asia are no longer a common occurrence.

In most cases, hunger in today's world is a function of food distribution and lack of economic purchasing power. The exceptions relate to droughts and refugee problems, which are discussed later.

There are some key observations related to the Green Revolution that bear on the discussion of the megawatershed potential. First, there would certainly not have been a Green Revolution in the 1960s was it not for the foresight of pioneers in the Ford and Rockefeller Foundations who funded the initial research stations as a public need and a potential public good. The individual researchers would never have achieved their breakthroughs without foundation help. The foundation role emphasizes the potential key role for public support for research to help accelerate development and adoption of new technology. In the case of early megawatershed research and development, the funding came from public and private sources such as USAID, British Petroleum, and private investors. More support is badly needed to help accelerate the adoption of the megawatershed concept on a broader scale, as we will discuss further at the end of this chapter.

Second, the Green Revolution technology was highly dependent on the availability of adequate, controlled water sources. The explosion in use of water for irrigation, highlighted in the IFPRI study, is partially due to this basic requirement. The Green Revolution, therefore, has greatly added to the pressure on the world's water resources. Much of the agriculture research in the large Consultative Group for International Research (CGIAR) network of research stations is focused on the problem of increasing food production in the face of water shortages. Dryland crop research on sorghum, millet, and other arid climate crops continues, but with far less promising results. Work is also being done with growing crops such as rice in brackish water conditions. The fact remains, however, that the major beneficiaries of the Green Revolution are concentrated in areas with adequate water resources, either from rainfall or water control systems. The application of the megawatershed principals may help make the application of new technologies, including biotechnology, feasible in arid regions (e.g. Somalia and Sudan) and erratic or heavily polluted rainfall runoff in high rainfall areas such as Trinidad and Tobago, providing year-round clean water for domestic use and food production.

Megawatersheds and Environmental Needs

There are a number of areas where the megawatershed concept and broad environmental concerns interface. We will touch on two of the key targets included in the UN Millennium Declaration, both of which are included under development goal No. 7, "Ensure Environmental Sustainability." The two targets are No. 9, "Integrate the principles of sustainable development . . . and reverse the loss of environmental resources" and No. 10, "Halve by 2015 the proportion of people without sustainable access to safe drinking water."

Target 9, megawatershed's contribution to preserving the natural environment, and the related issue of preserving biodiversity, are well illustrated in this book. In the case of Tobago, there was a determination by one of the world's leading engineering firms, using traditional hydrogeological methods, that there was no prospect of developing adequate water supply on Tobago through groundwater development. The only viable option presented was to dam the Richmond River in the Western Hemisphere's oldest protected natural rain forest. This park is the oldest natural preserve in the Caribbean and a source of important biodiversity. Developing the dam would have substantially impacted this reserve. An important related aspect is that the cost of the dam was substantial at $60 million. The development period was six years. The best water scientists in the world had spoken based on existing groundwater models and technologies.

Enter the megawatershed concept. As Chapter 5 demonstrates, the same quantity of water was developed and eight times more water was identified in one year than would have been possible even with full development of the dam in six to eight years, and the megawatershed development approach cost one-tenth as much as the dam with no adverse environmental impact. A bonus was the fact that the megawatershed approach allowed for private sector participation and financing. The dam as planned would have required all public financing.

In the context of this chapter, it is important to address Southern Africa, which was also extensively surveyed by Bisson and his team in the 1980s and 1990s, but it is not included as a full case study elsewhere in the book. The first regional deep groundwater study of Botswana and South Africa was undertaken by Bisson's firm BCI-Geonetics under contract with British Petroleum Southern Africa (BPSA) in 1985, followed by privately financed investigations of Zimbabwe and Mozambique and a USTDA-supported geophysical survey of the Sabe River Basin. In 1992, USAID engaged Bisson Exploration Services Company to assist the Agency in building a drought management strategy by following up on prior megawatershed's projects in Botswana. All three studies have regional implications for all of the SADCC states.

In these novel studies, attention was given to regional hydrological implications of underlying geological structure in Southern Africa as related to its major river systems, from the trans-African Zambezi to the Okavango and its remarkable desert delta. The Okavango Delta is one of the world's best known and loved wildlife preserves. Yet it is one of the few sources of fresh water in a country that is essentially a desert, albeit rich with diamonds. Many eyes are on the Okavango water. The delta catchment basin covers three countries: Angola, Namibia, and Botswana. Nearly 95% of the river inflows originate in Angola through the Okavango's two main tributaries, the Cubango and the Cuito rivers. Annual surface recharge to the Okavango Delta from Angola's Highland rains exceeds 10 billion m^3, with another 6 billion m^3 from local rainfall, all of which is channeled by the many faults and fracture systems that define the geographic expression of the Delta, from the arrow-straight and fault-confined Okavango panhandle, through the Delta to its truncating Thamalakane fault. Along the route from the Panhandle, 95% of that vast river recharge disappears from sight. Traditional hydrological models assume nearly all of that water evaporates, whereas the conclusion from Bisson's 1992 application of The Megawatershed Paradigm to the Okavango is that a large fraction of the annual Okavango flood contributes to groundwater recharge.

The megawatershed model presents an open-ended underground hydraulic system defined by the tectonically induced fracture fabric of Botswana and environs. Bisson postulated that the low-head, wave-like overland flow of the annual flood inexorably exerts downward pressure on the water table, activating a nine-month per year hydraulic pump of regional proportions. When new water is added to the top of that incompressible groundwater system, there are only two routes for the new water to take. If impermeable barriers surrounded the Delta's existing groundwater (the "bathtub" model), then the new incoming surface water would continue flowing overland, with far more water reaching Maun than is currently observed and with vast quantities of salt saturating the Delta's surface waters.

One current popular hypothesis is that 300,000 tons per year of solutes are captured by "Solute-Sink Islands," which, over thousands of years, have somehow collected and conducted hundreds of cubic kilometers of salts into deep storage in underlying aquifers. Although the fact that there are "solute sinks" scattered about the Delta is not surprising, the theory that tens of thousands of years of salt deposition amounting to thousands of cubic kilometers of accumulated salt can somehow be transferred and stored under the Okavango in such a manner seems implausible.

However, geological evidence clearly demonstrates that the groundwater of the Okavango Delta resides in highly permeable deltaic sediments underlain by wide-open, water-conducting fault and fracture systems that extend throughout the Delta and far out under the surrounding deserts. The megawatershed model, therefore, presents a hydraulically conductive Okavango groundwater system that responds to annual flooding according the laws of physics. As the flood surges add pressure to the top of the water table, the entire groundwater column is pressed downward by an amount equal to the pressure head added to the system at any given time (sometimes several meters). The incompressible groundwater pervading the Delta's conductive sands and bedrock fractures must then move along downward and outward along routes of least resistance, creating a low-head hydraulic network that follows bedrock fracture channels until directed upward by local bedrock structures to eventually evaporate (often without human detection) directly from solar-depressed subsurface water tables. In other venues, evaporating groundwater will leave its salt burden behind as evidence of its passing through.

Ironically, many saltpans in the Kalahari classified as lacustrine evaporates by hydrologists may be artefacts primarily of groundwater evaporation rather than surface water. Bisson posits that the Makgadikgadi and Sua Pans, which straddle a series of major, SW-striking faults could overlie massive groundwater discharge zones caused by fault-diverted upward water flows from regional deep groundwater within megawatersheds. The great distances from active recharge and huge salt quantities contained in these giant pans are quite feasible given the hydraulic power of a variable low-head (1–10 m), high-volume (8+ km^3 per year) hydraulic pump like the Okavango River.

It is clear that there will be competition for the water both in the source rivers and within the confines of the Delta itself. This is well discussed by Peter Ashton in his chapter of the United Nations' University publication, "International Waters in Southern Africa," published in 2003. The mining industry needs additional water. The capitol of Namibia, Windhoeck, also badly needs supplemental city water supply. Water is also short in other regions of Namibia. The rural populations along the Okavango River and around the Delta itself are already beginning to divert water, albeit in small quantities, for small-scale irrigation.

The mining industry has been eyeing the Delta itself. Windhoek is targeting the Okavango River that feeds the Delta. Clearly, there will be continuing efforts to tap into the Okavango water supply unless alternative sources are found. With the War ended in Angola, this will eventually include use of the resources of the Cuito and Cubango Rivers.

Based on the two Botswana megawatershed studies by Bisson (1985 and 1992), it is probable that adequate groundwater can be developed in the region using the megawatershed approach to meet the industrial needs of the mines, water supply for Windhoek, and other needs in Botswana without disturbing the ecology of the Okavango Delta. The second-stage water development phase proposed in the original study would be a major step in ultimately proving this potential. It is unfortunate that the studies have languished on the shelves for over a decade.

A third case example describes the megawatershed model as applied to the Sud in Southern Sudan. The traditional hydrological theory is that the White Nile flows into the Sud, overflowing its banks and flooding the area. The overflow is believed to have created this great wetlands area, the second largest wetland in Africa. French engineers began a major dredging of the White Nile in 1978 to create the Jonglei canal, expecting the canal to enhance the flow of the Nile. The current overflow of the Nile would then be reduced along with the associated evaporation losses in the Sud; it would therefore be possible to recapture water that was formerly "lost" in the Sud. It was estimated that the first phase of the project would boost the annual flow of the White Nile by $4\,km^3$ and the second phase by an additional $4\,km^3$. This recaptured water would then be used for the benefit of downstream Sudan and Egypt.

Given the history of the Nile and its importance to Sudan and Egypt, this was, at the time, an appealing concept to both countries, which agreed to share the costs and benefits of such projects in the 1959 Nile Waters Agreement [Allen and Tauris, 2002]. The 260-km canal would also have potential transport benefits by substantially reducing the distance through the Sud. However, in retrospect, it may be that the only constructive outcome of the horrendous war in southern Sudan in 1984 is that the dredging had to be abandoned, permitting the development of a new, comprehensive hydrogeological model that was applied in a 1987 USAID contractor's study.

The study, carried out by BCI-Genomics, Inc. (Bisson, et al. 1987) reported that the Sud was formed by groundwater discharge caused by an intersection of two major bedrock structural systems related to Nile and Red Sea tectonics. The bedrock underlying the Sud is shattered, creating a very large area of high fracture permeability. These open fracture systems extend far into the Ethiopian highlands, where high rainfall and elevation combine to produce a high-volume, high-head hydraulic pump that efficiently carries large amounts of groundwater along fracture zones westward into the Sud, where groundwater flows are diverted upward to surface by North-striking faults. Bisson postulated that more than 30% of rainfall in the Ethiopian highlands could recharge fracture zones that eventually discharge into the Sud.

The Sud megawatershed model indicates that the White Nile is not the controlling hydrological element in local hydrological balances. In fact, historical evidence indicates that the Sud water level is at times higher than the White Nile, causing reversal of the White Nile's flow. The hydrological and hydraulic models used by the French engineers may be faulty. Renewing the dredging project may risk doing irreparable damage to one of Africa's major natural wetland resources while a major water resource goes unutilized. Detailed megawatershed studies related to the Sud have not been done. There may be some approaching urgency to do such an analysis.

The war in southern Sudan is hopefully coming to an end. There will certainly be consideration to revisiting the Jonglei project. John Garaang, the leader of the southern resistance, in fact, did his thesis in the United States at Iowa State University on the Jonglei canal. Certainly, after the war, there should be a more thorough reassessment of the hydraulic model and the impact of the project before any attempt is made to resurrect the old plan, or any new variation. J.A. Allen, in his discussion of the project, argues that increased environmental criteria adopted in the last 15 years will make the resumption improbable. Sandra Postel (1993) postulates that the southern Sudanese would oppose any resumption in the near future. In the long term, pressures in both Sudan and Egypt for increased water availability from the Nile will only increase. Unless an alternative framework is put in place, the pressure to dredge the canal will continue.

The implications of the Sud megawatershed are enormous. Untapped renewable groundwater resources may amount to cubic kilometers per day, with a large fraction flowing beyond and around the Sud, to pass beneath villages and farms throughout the region and eventually evaporate in a desert wadi. Large amounts of this surplus groundwater could be safely managed for beneficial use throughout the region with no measurable effects on the Sud or the Nile's flow.

These are but three illustrative examples of how introduction of the megawatershed model can contribute to preservation of the world's natural resource environment and preservation of biodiversity.

The second environmental question is how megawatersheds might contribute to the UN Millennium goal of improving sustainable access to safe drinking water.

It is estimated that 1.1 billion people worldwide lack access to clean water and 2.4 billion lack adequate sanitation. As the study "Financing

Water for All" [World Water Council, 2003] states, "Poor water and sanitation is an important cause of diseases such as diarrhea (4 billion cases, with 2.2 million deaths), plus intestinal worms, cholera, Schistosomiasis and other water related health problems." Improved water supply is not the only answer, but it is an important part of the solution. This study indicates that the investment required for drinking water supply is 13 times the investment for improved sanitation and hygiene. So the need for water supply is critical. The reality for providing it is discouraging. For example, the study points out that the United States, Australia, and Ethiopia all have similar climates, but Australia and the United States have $5000\,m^3$ of water storage per person, whereas Ethiopia has $50\,m^3$. Why is this the case? Preliminary studies of the megawatershed potential in and around Ethiopia have been carried out under USAID contracts. These studies indicate abundant additional groundwater potential (e.g. the Sudan and Somalia case studies in Chapters 3 and 4). In addition, the USAID OFDA study on Botswana referred to above includes a parallel assessment of water potential in areas of Ethiopia itself. There appears to be substantial potential for tapping groundwater to meet the needs of the Ethiopian people for drought relief, drinking water, and possibly agriculture needs as well. This will be further discussed in the section on droughts and refugees.

It is clear that if the world is to meet the drinking water targets, especially in areas like Ethiopia, then greater consideration needs to be given to application of the megawatershed concept along with more traditional approaches. This could open new and potentially costly water sources. In addition, using megawatershed models, governments can "act locally" and institutions can "think globally" because once drilled the wells can be strategically sited and locally managed. It might, therefore, be possible to avoid some of the major management problems associated with traditional centrally owned and managed water systems. Megawatersheds thus present an interesting prospect for greater use of locally owned and managed systems once the major countrywide development work has been completed. One of the major challenges today in the developing world is the financing and management of centrally managed water systems. Increased reliance on locally managed systems helps meet this challenge.

Megawatersheds and the Water Needs of Refugees and Drought Victims

Prior megawatershed projects in Africa have highlighted the important role that this technology can play in relieving the life-threatening conditions that repeatedly threaten refugees and drought victims. This is most clearly illustrated in the case study on Somalia where the refugees were given access to water through the USAID-funded project in 1984–1986. Unfortunately, the civil war pushed many of the refugees over the border into the Ogaden. The project was halted before full development. Decades later, these wells are still operating successfully. Despite its early termination, the Somalia project highlights the critical role that megawatershed technology can play in refugee/resettlement programs.

In 1989, a few years after the Somalia project, it was noted that large expenditures were being incurred by the UN refugee relief agencies to truck drinking water down the mountain to the refugees in the Ogaden. The closest Ogaden refugee camp to Somalia was very close to the location of the wells drilled by BCI in Somalia. BCI estimated that an adequate water supply could be developed using the data developed in Somalia to meet the needs of the Ogaden refugees. The total cost to drill the wells would be less than the annual cost of trucking water. The proposal was forwarded to the UN by the State Department. It disappeared into the UN water bureaucracy a casualty of the civilian and another missed opportunity.

Credit goes to USAID, however, which early on supported testing of these novel technologies and which later realized the potential of megawatershed technology to meet the need of both refugees and drought victims in Africa. Refugees and droughts are not a one-time event in Africa. They are an unfortunate part of the fabric of Africa's underdevelopment: its flawed political leadership, in certain countries, and a result of arid conditions throughout the Sahel, the Horn, and in Southern Africa.

In 1992, The Office of Disaster Assistance (OFDA) of USAID did seek to follow through on earlier studies. USAID commissioned a perceptive prefeasibility to examine the application of megawatershed principles to refugee problems in Ethiopia and drought conditions in Botswana. The Botswana study was discussed in the above section on the environment, but it was primarily directed on Botswana's continuing drought problems. As we note above, the results indicated substantial potential for groundwater development to meet future drought needs. The next stage of work still waits to be done.

The study in Ethiopia was also encouraging regarding the potential for groundwater development in both Tigray and Eritrea. Today the refuge problems remain and are compounded by another horrible drought in Ethiopia. Droughts have become a repeat occurrence in Ethiopia with the populations facing new stress before fully recovering from the last crisis. The review by USAID in 1992 indicated that development of megawatersheds in Ethiopia can respond to both the country's drought and refugee-related water needs as well as contribute to long-term development in the rural areas.

In addition, there is growing recognition that there must be better links between drought relief and long-term development strategies. A more dependable source of water supply would help both emergency and long-term development programs in Ethiopia. Unfortunately, the encouraging prefeasibility work in Ethiopia has also not yet proceeded to the next development phase. In fact, the study seems to have been forgotten. The original study in USAID's central database is unavailable. The exploratory drilling phase of the work might have been completed in the 1990s. This would have helped during the current drought crisis, which is threatening the lives of millions of Ethiopians once again.

A final observation on drought mitigation, a Famine Early Warning System (FEWS), was established following the African drought in 1984. The program supported by the United States does sophisticated weather predictions associated with assessments on the impact of weather on crop production and food supplies. Reports and special analyses are done for both African countries and International donors. The program has been successful and is being expanded to Afghanistan and possibly other countries and regions. However, despite its success, imagine how much more powerful the FEWS assessment process would be if a fuller understanding of groundwater sources were added to the assessment process. It would be literally adding a new dimension, to the program. This program would allow projection of where groundwater might be used to meet drought needs when rainfall fails and surface water and shallow aquifers dry up.

Megawatersheds are, of course, not the sole or the primary answer for solving Africa's drought and refugees' problems. Improved political leadership and more effective economic development are basic requirements for a long-term solution. But megawatersheds, and the provision of new sources of water, will help meet emergency needs until that day. Even then, as in the case of Botswana where there is excellent leadership and good progress in economic development, there will still be a major role for megawatersheds in dealing with drought.

Megawatersheds and Commercial/Industrial Use

Much of the discussion in this paper concerns priority needs related to drinking water and water for agriculture production. Water is, of course, also essential to the undertaking of a wide variety of commercial and industrial applications. In this book, there are a number of examples of how additional water, located through megawatershed techniques, can aid business.

In the case of Tobago, a major hotel complex was completed with the expectation that adequate water would be available for both the hotel and related golf courses. When the complex was completed, there was no water, and the multimillion-dollar hotel complex sat idle for over a year. The successful location of groundwater by Earthwater Technology and Lennox Petroleum of Trinidad effectively saved this US$100 million development project from potential failure.

The Sudan case study presents a much more complex model for using megawatershed analysis to launch a major new regional development scheme. The proposed development envisioned a broad range of commercial and industrial projects proposed by private Sudanese investors. The development centered on the vision of restoring the ancient Port of Suakin. A new port complex, hotels, and a variety of commercial activities were studied and found viable *if* there was adequate new water supply. The associated megawatershed analysis indicated sufficient water was available. The project was stopped for political reasons not economic or technical ones.

It is clear that throughout the areas supplied with supplemental groundwater, there will be a wide variety of new opportunities for commercial development that did not earlier exist. Many of these will be small-scale enterprises with a large social and economic impact on relieving poverty. This will be especially true in rural areas where commercial development will go hand in hand with increased agriculture development also facilitated by groundwater derived thanks to the megawatershed model.

Megawatersheds and International Political Challenges

At the international level, the shortage of water, especially in certain areas, such as the Middle East, makes water one of the world's key political challenges. Lasting peace will require lasting solutions to competition over water shortages. Outside the Middle East, the issue of competition for water is also acute. As Sandra Postel highlights in her 1993 article in World Watch on the Politics of Water, "nearly 40% of the world's population depend on river systems shared by two or more countries." We have already touched on two of these systems. The first is the Nile shared by nine countries. The second is the Okavango River shared by three countries. In both cases, we have discussed how megawatershed technology might contribute to water challenges of interest to the affected countries, e.g., the Sud and the Okavango Delta. In both cases, the principle involved is the introduction of a new water source to an equation that concentrated almost solely on allocating surface water. Surface water is, of course, at the root of the debate in the Middle East, although shallow aquifers also are brought into the equation.

One can ask, as many already have, what might the contribution of additional groundwater, developed through megawatershed technology, add to solving the water problems in the Middle East. It is clear that a substantial volume of rain falls in the highlands of Turkey. This probably contributes to substantial groundwater recharge through the known porous geology of the region. The stories of leakage in the reservoir of the Keban Dam, during construction, highlight this fact. Why then has there not been a greater effort to explore the possibility of additional groundwater in the Middle East? Alternatively, more traditional surface options have been proposed. Turkey once proposed two major water pipelines at a cost of $21 billion [Postel, 1993].

The concept of adding a new groundwater dimension to the water debates has global significance and applications. An additional example is in the Central Asia region where five former Soviet Republics compete for water from the Amu Darya and Syr Darya rivers. Major Soviet irrigation schemes tapped these rivers and left a legacy of high-maintenance irrigation systems. Yet, in this area, there are many manifestations of substantial groundwater resources. The region is also one of active tectonics, faulting, and heavy rainfall conducive to large megawatershed potential. Here, the addition of major untapped groundwater offers substantial potential for reviving the agriculture of a region that has great similarities to the Central Valley of California.

In summary, there has been abundant attention given to the political implications of competition for scarce water sources throughout the world. In almost all cases, this debate takes place in a two dimensional world focusing primarily on surface water resources. If we are willing to add an expanded third dimension of major new groundwater resources through application of the megawatershed concept, the options for solving the political conflict increase exponentially.

Megawatersheds, Skepticism, and Risk Reduction

If megawatershed technology offers so much promise, why has it not been more widely applied? As the man said, "if this so important, why is it not happening?" The first barrier is outright skepticism blended with a fair dose of vested interest in the status quo. This fact is epitomized by the story of Max Planck's professor who was tied into the conventional wisdom about physics in his day. The skepticism is also reflected in the von Helmholtz quote starting this chapter. A major challenge faced by the proponents of the megawatershed theory is to overcome the resistance in the professional water community. More successful case examples like Trinidad/Tobago will help overcome skepticism. There is also a need to chronicle other successes or potential successes already studied, e.g., Sudan, Botswana, and Ethiopia.

This approach will eventually do the job, but it is a very slow process. It is nevertheless an essential part of the learning process. Groundwater applications need to be accelerated. There is a bit of the chicken and egg syndrome here. Because of skepticism, it is difficult getting support for new on-the-ground applications. Without more on the ground successes, the skepticism persists.

In a world of continuing skepticism, the perception of risk associated with financing a megawatershed initiative project is high. This barrier to support could, however, be reduced by techniques of reallocating the risks. The challenge is to make the risk-benefit equation more attractive to project owners and financiers. First, water projects worldwide, especially municipal water systems, are predominately publicly owned or financed. The report of the world panel on financing water infrastructure "Financing Water for All" highlights the basic character of the water sector. "Water and Sanitation financed 65 to 70% by the domestic public sector, 10–15% by international donors, 10–15% by international private donors and 5% by local private sector." International donors and public agencies mostly fund major irrigation and drainage systems, including large dams. Smaller systems depend on local finance.

This profile highlights the fact that much water development is overseen by bureaucrats in the domestic public agencies and with the international aid donors. Private international and domestic investment is important, but not dominant. I am a former public bureaucrat myself. I know that the risk-benefit equation in the public sector tends to punish failure more than it rewards success. The opposite is true in the private sector. Dealing with these facts is critical to winning support for additional new megawatershed projects. Let's take a look at how the risk-benefit has been shifted between the public and private sector so far.

Two of the earliest examples of providing additional water supply from unprecedented bedrock sources using novel business vehicles were in the New England towns of Seabrook and Salem, New Hampshire in 1980–1981. The seacoast town of Seabrook had previously drilled over 150 test wells in an unsuccessful search for fresh water in alluvial deposits and in the bedrock and was then considering surface water diversion and a desalinization plant to meet the city's critical water supply needs. Inland 50 miles, Salem's Canobie Lake water source was drying up, and a very expensive pipeline to the Merrimack River and treatment plant were being contemplated.

BCI Geonetics, Inc., which pioneered fractured bedrock aquifer studies, approached the towns with the deep bedrock aquifer alternative, offering to identify and develop new water sources to meet cities' needs at far lower cost than surface water or desalination. City supervisors were skeptical that anyone could drill holes in good old New England "granite" and extract water.

Seabrook decided to sign a standard "time and materials" contract and was rewarded with more than 1.5 mg of potable water within a year of starting the project at a total cost lower than the fees for project design alone would have been for either of the other alternatives.

Salem would not risk city funds in such a venture, so BCI prepared a revised proposal that eliminated the contract risk by agreeing to simply sell them water at a guaranteed per-gallon price over a multi-year contract period. Salem signed the contract. BCI brought in investors and a drilling partner and proceeded to explore for and develop over 1.5 million gallons per day of portable water from the highest yield "hard-rock" wells ever drilled in New Hampshire, some of the walls which produced 100 times more water per day than a normal bedrock well.

What happened here? The city supervisors did not lose their skepticism through the contracting process. Rather, the risk was shifted from the city to BCI. BCI and their investors assumed additional risk and recovered their investment through successful development of the wells and subsequent water sales.

Twenty years later in Tobago, BCI's founder, Robert Bisson, and some of his original exploration team members, operating under the aegis of Earthwater Technology, combined a fully evolved megawatersheds' exploration technology and team knowledge base with a newer, more advanced public-private sector risk-sharing model to successfully solve a 45-year water shortage on the Caribbean island of Tobago. This contract model was documented in a paper prepared by Tobago's senior water authority management for the tenth annual conference of the Caribbean Water and Wastewater Association (CWWA). The paper was entitled "High-risk Groundwater Development Options for Small Island Developing States" (SIDS). The director of the Trinidad Tobago Water Resources Agency (WRA) presented a second paper on the subject to the fourth Inter-American Dialogue on Water Management in Parana, Brazil in 2001.

The key ingredient of the Tobago approach was the separation of the contract into two risk categories: the first phase hydrogeological mapping and the second stage drilling of the development wells. The cost of the first phase mapping effort was borne by the Water and Sewerage Authority of Trinidad and Tobago (WASA) on a lump-sum contract award. Successful completion of this phase provided WASA with a new, comprehensive mapping of the island's hydrogeology. Secondly, the completion of this phase allowed the contractor to better assess the risks and potential for successful development of new wells. This in turn allowed the contractor to reduce the risk associated with well development. This reduction in risk ultimately benefited both parties.

In the drilling phase, the drilling contractor was to be paid only on a successful outcome. This meant that if the contractor did not find and prove predetermined quantities of high-quality water, the vendor would not get paid. WASA, therefore, bore essentially no risk at the drilling stage. As the case discussion shows, the Tobago effort was very successful and led to a follow-up larger project in Trinidad itself. The Trinidad project was also successful and followed the same risk-sharing model as in Tobago.

In both the Seabrook project and the Trinidad/Tobago projects, the drilling principle was that no payment was made unless water was found in adequate quantity and quality. This differs from common drilling contracts for water where the contractor is paid for a level of effort usually on a time and materials basis. A contractor drills the well and gets paid, whether water is found or not. This approach is therefore new to water well drilling.

The performance payment principle is not new to oil companies. A closely related case in geothermal development shows the parallel. In the 1970s, the Philippines was a country with an energy crisis. The Philippines has no oil, no gas, and few coal reserves. However, the Philippin does have good hydroelectric, confined to only two out of the 7000-plus islands. So the Philippines committed them to a major nuclear plant. A request to USAID for support in a small geothermal project at the same time generated opinions from the technical experts that geothermal technology was unproven and risky. This reflected comments also received from UN "experts" despite the fact that there were successful geothermal developments in Italy, New Zealand, and Iceland and in the Geysers in Northern California. A subsequent dialogue took place among Philippine government officials, Union Oil's geothermal division (at the Geysers), and Pacific Gas and Electric Co., purchaser of the steam at the Geysers. This exchange led to an investment contract to develop the TIWI geothermal field in Southern Luzon. Union Oil took the drilling risk against a commitment by the

Philippine government that they would buy the steam and build a thermal plant to use the steam. Today the Philippines has a substantial geothermal development program, including on islands with no other indigenous power source. Geothermal-fueled electricity far exceeds the capacity of the hugely expensive Nuclear Plant, which has never operated because of earthquake risks.

These are but three examples of where cooperative risk sharing between the public and private sectors allowed successful development with a new technology perceived to be of high risk.

THE WAY AHEAD

As described above, there are several steps necessary to promote greater utilization of megawatershed technology. There is a key role that the public sector bodies and the international donors might play. There is a critical role beyond the financing of individual projects, although that is also important. What is needed is a much more systematic effort to develop a far more accurate hydrogeological profile of major sections of this planet using the technology associated with the megawatershed paradigm approach. Great strides have been made in the fields of remote sensing, and geographic and geophysical technologies that allow for such an effort at far lower costs than in the past. Such an investment in what is truly a "public good" would yield incalculable benefits in future development of the world's usable or portable water supply. The financial costs of water development and the risks to private developers and public bodies would be greatly reduced. New horizons would open up. New solutions to the water challenge would appear.

Helmholtz would smile down on us all.

BIOGRAPHY

Alexander R. Love is founding Executive Director of the Washington DC-based Partnership to Cut Hunger and Poverty in Africa, a non-profit organization dedicated to assisting Africans in reducing hunger and poverty. He serves on the Partnership's executive committee and chairs the subcommittees on AIDS/Agriculture and Infrastructure.

Mr. Love's four-decade career has focused on multilateral and bilateral development assistance programs for third-world nations, including infrastructure and technical programs directed at increasing food production, improving health, and reducing poverty. He was instrumental in developing and implementing USAID and OECD policies, strategies, and programs for development assistance worldwide. His particular interests are in sub-Saharan Africa, East and South Asia, where he developed and directed numerous USAID programs, including many contributing to the successful "Green Revolution" initiative. Mr. Love served in capacities ranging from senior positions in project finance to senior management positions such as Director of the Regional office in East and Southern Africa, Deputy Assistant Administrator for Africa, and Assistant to the Administrator for Personnel and Finance.

Mr. Love served as a consultant to the Board of the World Bank group (2003) and was appointed by the Secretary General of the United Nations to the Panel of Eminent Personalities (MEP) charged with evaluating the success of the UN New Agenda for Development of Africa (1990–1991). Mr. Love was Chairman of the Development Assistance Committee of the OECD in Paris (1991–94), where he coordinated policies and strategies for administering over $60 billion in development assistance from twenty-one donor countries.

During his tenure at USAID, Mr. Love was appointed Counselor to the Agency (1987) (the career deputy to the administrator). He received the Presidential Superior Honor Award, the Presidential Meritorious Honor Award, and the Luther H. Replogle award for management excellence in administering drought assistance during the 1984 Africa drought.

Mr. Love holds undergraduate degrees in civil engineering and business administration from UC Berkeley and a Masters degree in business administration from Harvard University.

REFERENCES

Allan, A.J., and Tauris, I.B., 2002, The Middle East Water Question/Hydropolitics and the Global Economy.

Ashton, P., 2003, International Waters in Southern Africa, in Nakayama M. (Ed.): The Search for an Equitable Basis for Water Sharing in the Okavango River Basin, United Nations University.

Bisson, R., Hoag, R. Jr., et al., 1985, Report on Regional Groundwater Occurrence and Economic Development for Botswana, South Africa, and Southwest Africa: Developed for BCI Geonetics, Inc. and British Petroleum, South Africa.

International Food Policy Institute, 2002, World Water and Food to 2025/ Dealing with Scarcity.

Maharaj, U.S., Bisson, R., and Hoag, R. Jr., 2001, In Quest of Solutions.

Maharaj, U.S., and Tota-Maharaj, T., High Risk Groundwater Development Option for Small Island Developing States.

Postel, S., 1993, The Politics of Water: World Watch Magazine.

The Partnership to Cut Hunger and Poverty in Africa, 2002, Now is the Time/ A Plan to Cut Hunger and Poverty in Africa: Strategic Plan.

World Water Council, March 2003, Financing Water For All: Report of the World Panel on Financing Infrastructure: 3rd World Water Forum, Global Water Partnership.

3 Case Study—Northwest Somalia 1984–1986*

INTRODUCTION

The Somalia regional fractured bedrock aquifer study covered an area larger than the state of New Hampshire and was unprecedented in its goals, hydrogeological concepts, technical methods, and groundwater discoveries. This 1984–1986 study launched a new era in groundwater hydrology.

BACKGROUND AND SETTING

In 1984, the Horn of Africa was experiencing a particularly brutal famine, brought about by the coincidence of periodic drought and regional and civil wars. America's ally, Somalia, was at war with the Soviet Union's ally, Ethiopia. At the same time, Somalia was mired in a bloody civil war, and the most active front for both conflicts was the remote, sparsely populated northwest part of Somalia, 1500 road-kilometers from Mogadishu, bordered on the west and south sides by Ethiopia, to the north by the Gulf of Aden and the northwest the last bastion of French colonialism in Africa, Djibouti. Several hundred thousand ethnic Somali refugees had been forced from their ancient tribal lands (annexed to Ethiopia after WWII), in the ongoing war and sought refuge in the arid Woqooyi Galbeed and Tog Dheer provinces

* Summarized from source material: Technical Report of Groundwater Exploration in Northwest Somalia for Refugee Self-Reliance and Productivity: Regional Groundwater Resources and Test Drilling Results; by R.A. Bisson, R.B. Hoag, F. El-Baz, and others for BCI-Geonetics, Inc., 1986; under contract with New Transcentury Foundation–USAID Cooperative Agreement.

of Somalia (Figure 1). As a result of these circumstances, the socioeconomic fabric of the region was destabilized, cholera and other diseases were rampant in refugee camps located on the outskirts of major urban centers, and virtually no food was being produced locally.

Somalia's new patron, the United States, was working with the UN High Commission for Refugees (UNHCR) and the United States Agency for International Development (USAID) to help feed and shelter those desperate people in an arid, undeveloped part of the Horn of Africa with all possible speed. USAID had contracted with New Transcentury Foundation (NTF), a nongovernment organization (NGO) based in Washington, D.C., to provide emergency road construction and water resources development for the refugees under an innovative plan aimed at stabilizing the region while assisting refugees, called the "Refugee Self-Reliance Program," stabilizing the region's human populations by enabling ethnic Somali refugees to become self-sufficient in water and food production in the next growing season in previously undevelopable and uncontended areas, rather than on the periphery of urban areas, where many refugees were congregating and creating imminent risks to public health and tranquility.

Early in 1984, Warren Wiggins, president of NTF and a U.S. foreign aid veteran revered as the principal architect of Kennedy's Peace Corps initiative, contacted Robert Bisson, president of BCI Geonetics International, Inc. (BCI). Wiggins requested that Bisson visit Somalia to assess the potential for additional groundwater development and submit a plan to identify the groundwater resource base of the region and develop emergency water sources and long-term water supplies for new refugee encampments. Bisson's field assessment led to BCI initiating a regional groundwater exploration and test-drilling program in 1984 under subcontract with NTF that covered an area of approximately 35,000 km².

GENERAL DESCRIPTION OF STUDY AREA

The Study Area was located in the Woqooyi Galbeed and Togdheer Provinces of northwest Somalia, extending from the international borders shared with Djibouti and Ethiopia to 45°E longitude. The population was concentrated in two cities and several refugee camps. Hargeisa, the regional capital and largest city in the area, was the location of the NTF field office, which served as a base for BCI's field operations. Borama, a smaller settlement located to the west of Hargeisa, was on the Ethiopian border. The rest of the primarily nomadic population is scattered throughout small villages and temporary encampments.

Only one paved road system existed in this region, connecting Hargeisa with the port city of Berbera and extending 70 km west toward Borama. A limited network of mostly unimproved dirt tracks that become impassable during the rainy seasons connected all other settlements. Topography varies considerably with location and is primarily a function of the underlying bedrock geology and structure. Sites visited in the crystalline metasediments of the inland crystalline highlands such as Agabar, Waragadighta, and the Jire Wadi are generally characterized by rugged terrain and steep weathered slopes, whereas areas primarily underlain by limestone and sandstone, including Hargeisa and Borama, exhibit a more gently rolling topography. Access to all sites evaluated in this report is possible during the dry seasons with a four-wheel-drive vehicle (with the exception of Shirwa Basati, which presently must be accessed by foot) via either wadi beds or existing tracks.

With the exception of some small farms along the wadi beds where water is readily available and some occasional plots of rain fed crops, very little land was under cultivation. Most of the region is sparsely vegetated and flat, and it was used as rangeland for grazing herds of camels, cattle, sheet, and goats. Along the wider wadis, plant life was more diversified and lush and many of the larger tree species flourish.

During the course of geophysical investigations July–August 1985, the climate was seasonably hot and dry, with daytime temperatures between 24°C in high inland areas like Borama to 46°C at the Biji Valley site where the Biji Wadi exits the mountains onto the coastal plain. Some rainfall was encountered in the central uplands of the Dibrawein Valley and Hargeisa, and storms also passed through the hills in the Biji and Durdur Wadi watershed during this period.

EXECUTIVE SUMMARY OF PROJECT GOALS AND RESULTS

Previous water development studies of NW Somalia by other investigators focused on surface and surficial water sources and either ignored the potential of bedrock water resources or briefly mentioned possible development without quantifying potential. BCI's team focused on the region's bedrock groundwater potential, and the team investigated groundwater recharge and storage in unconsolidated overburden only where they represented integral hydrological parts of deeper, underlying bedrock aquifer systems.

The project had three goals, including (1) generating a first-ever strategic groundwater development map for the region, (2) providing refugees with near-term water from fractured bedrock aquifers using test wells, and (3) drilling of high-yield deep bedrock production wells for long-term agriculture and municipal use.

In spite of the Force Majeure termination of the project early in Phase 3, the team achieved critical goals, including (1) a detailed exploration report and maps identifying potential untapped deep groundwater resources exceeding 20 million gallons per day over the Study Area, (2) nearly 1.5 million gallons per day of high-quality groundwater in six test wells located near refugees and indigenous populations, and (3) verification of the crystalline bedrock groundwater model BCI used to create the strategic groundwater development map. Some of BCI's test wells reportedly have continued to supply fresh water to Borama's citizens and others for the past 16 years [Lisa Wiggins, personal communication].

WATER DEVELOPMENT TEAM

The NTF project director, Lisa Wiggins, provided overall coordination of the emergency road construction and water resources aspects of NTF's contract, and the water exploration team comprised BCI's team leader Robert Bisson and senior advisors Dr. Farouk El-Baz and Dr. Lincoln Page as well as BCI Geonetics explorationists Roland Hoag, Joseph Ingari, Rosemarie DeMars, Dennis Albaugh, Jamie Emery, and an interdisciplinary team of senior scientists from BCI Geonetics, Itek Corporation, and Earth Satellite Corporation.

LOGISTICAL CONSTRAINTS AND CHALLENGES

In order to carry out their hydrogeological mapping mission, the BCI field team found few contemporary cultural and land-use maps and so augmented older maps with a combination of 1950s-vintage 1:125,000 scale British Ordinance Survey "spot-elevation" maps and RAF aerial photographs, plus 1980s-Tactical Pilotage Charts (TPCs), LandSat Image-maps, and high-resolution Space-Shuttle-borne Large Format Camera (LFC) photographs. The aerial and LFC photos were very useful navigation tools in certain remote areas where there were no roads leading to key bedrock outcrops and other sites identified by BCI scientists on aerial photographs and satellite images. Thorn-bearing plants and sharp rocks punctured tires and automobile anatomies with great regularity, and undercarriage-armored Land Cruisers important via neighboring Djibouti were outfitted with 12-ply tires for team expeditions.

As the BCI team was tasked to explore the farthest reaches of the county's lawless northwest region, they often penetrated trackless areas where no vehicle or camel track existed and crossed unmarked borders between Ethiopia and Somalia. They were in constant danger of being captured by Ethiopian troops, strafed by MIGS during daylight and held up at unmarked roadblocks by Somali bandits, revolutionaries and government soldiers at night. The region's wilderness also abounded with non-human predators, from hyenas to cobras, poisonous centipedes and the nightmarish platter-sized, venomous "Camel Spider". The BCI team encountered in Somalia the most dangerous and difficult field conditions they had encountered to date, including one memorable confrontation with Somali soldiers in a remote area near the Ethiopian border that very nearly cost the lives of BCI's core team.

After completing the groundwater-mapping program and commencing the test drilling, the BCI team was pulled out of Somalia by USAID in 1996, only several months and six test wells into the test-drilling program. The six test wells yielded a noteworthy 1.5 mgd. The sudden termination was due to a radical increase in expatriate kidnappings by rebels and impending invasion of the northwest by Somalia's central government army, bent on destroying the insurgents. Soon after the team's departure, Somali army tanks and artillery flattened the team's home base in the regional capital city of Hargeisa, and local Somali NTF and BCI team members were murdered during the conflict.

TECHNICAL APPROACH—OVERVIEW OF MESA© PROGRAM APPLICATION

BCI Team Technical Approach

The BCI team conducted this investigation using its mineral-cum-groundwater exploration method, the Mineral Exploration Systems Analysis (MESA©) Program. This novel exploration program, developed by BCI's founders and key scientists through extensive trial-and-error field testing in the decade prior to the Somalia project involves a systematic, practical approach to quantifying targeted minerals, including deep groundwater occurrence, over large areas. The MESA© Program was found to be singularly successful at identifying, drilling, and developing sustainable groundwater sources nearby consumers in the most economical way possible.

The premise of the MESA© Program is groundwater shares and is controlled by the same geological environments as minerals and is most readily identified and evaluated using modern approaches proven successful in oil, gas, and economic mineral exploration. The Program applies state-of-the-art technologies and then integrates and evaluates background data, remote-sensing interpretation, structural geologic mapping, surficial geologic mapping, inventory of existing boreholes for quantity and quality information, identification of sources of contamination, aquifer management, and protection.

The successful application of this sophisticated Program requires a practiced exploration team of experts in structural geology, remote sensing, surficial geology, exploration geophysics, geochemistry, hydrology, and hydrogeology. Only with an integrated team such as this can the full groundwater potential of a Study Area be determined.

In 1984 the BCI-Geonetics MESA© program consisted of six phases with a predetermined milestone reached at the conclusion of each. At the milestones of Phases 1 through 4, the probability of success in certain parts of the study area increased, while other parts of the study area were eliminated from further consideration. The six phases were:

Phase 1: Favorable zones identification, including background data collection, remote sensing data acquisition and interpretation, structural geologic mapping, surficial geologic mapping, inventory of existing boreholes for groundwater quantity and quality information, identification of existing sources of contamination (both natural and cultural) and determination of groundwater recharge and discharge.

Phase 2: Reconnaissance-level geophysical and geological surveys to identify most favorable water development target areas within favorable zones defined during Phase 1.

Phase 3: Detailed geophysical surveys to site test wells in most favorable areas.

Phase 4: Test drilling.

Phase 5: Production well drilling and well development (production enhancement)

Phase 6: Pumping facilities construction.

MESA © PROGRAM APPLICATION IN SOMALIA

Discussion of Methods and Outcomes

The original scope of the Somalia investigation encompassed MESA© Program Phases 1–6; however, the evacuation of the BCI team by USAID in 1986 confined the scope to partial completion of Phase 3 test drilling. The 1.5+ million gallons per day of new potable water from just six test wells represent a small fraction of the potential sustainable resources identified in northwest Somalia during the groundwater exploration phase of the MESA© Program.

Surface waters and shallow aquifers had previously been studied by other investigators, and they were developed to economically practicable capacities. The BCI study was primarily focused on bedrock aquifers, based on previous work by BCI investigators in similar geological terranes. The BCI team initially divided the 35,000-km^2 groundwater exploration Study Area into three major hydrogeological provinces, including the coastal plain and piedmont, the inland crystalline highlands, and the mid-Tertiary plateau, with aquifer types distinguished by predominant bedrock lithologies. Certain areas with least hydrogeologically favorable bedrock were eliminated from in-depth study to achieve the time-sensitive, problem-solving goals of this venture. Within hydrogeological provinces, the primary geographical unit used to compile hydrological data was the fourth-order drainage basin. Each fourth-order basin in the Study Area was numbered, and the general hydrogeological characteristics of each basin was tabulated.

The hydrogeological provinces were further screened using a criteria matrix containing certain key geologic attributes known by the team to be relevant to the groundwater exploration program. Each attribute was assigned a "weight," on a scale of 0–10, equal to the level of importance of that factor in the conceptual model being applied. This preliminary analysis resulted in identification and a hierarchical rating by groundwater potential of numerous subareas within the hydrogeological provinces, with each area ranked from highly favorable to unfavorable.

Highly favorable subareas, called "Favorability Zones," were then evaluated for their hydrological characteristics (active groundwater recharge, storage, and transmission), and only the highest-rated Favorability Zones were selected for field investigations within the scope of this program. Lower-ranked Favorability Zones and numerous other areas also exhibited potential for future groundwater investigations.

Somalia Technical Program: Summary of Methods and Results

Favorable Zones Identification   Before initiating new fieldwork, the team thoroughly evaluated the data and conclusions of previous investigators. The team found a substantial body of work performed by previous investigators who had applied conventional hydrogeological theory inappropriate for northwest Somalia. The old theory did not recognize new hydrogeological paradigms that described the occurrence of actively recharged deep bedrock aquifers in complex geological terrains exemplified by the Somalia Study Area. Others attempted to site water wells in bedrock fractures zones without a conceptual model to provide critical information about aquifer recharge, storage, or productivity.

For example, basic "fracture-trace" maps produced by Christiansen et al. [1975] and Kronberg et al. [1975] were available for the northwest portion of Somalia and the Afar Depression. The fracture-trace data consist of transparent Mylar sheets covered with thousands of straight and curved lines traced by hand from linear features observed on satellite images by photoanalysts. Unfortunately, the "fracture-trace" method of water well siting, first reported by Lattman and Parizek in 1964, often improves average well yields only in relatively uncomplicated carbonate rock terrains. Therefore, as the prior "fracture trace" image analyses in Somalia were not performed in context with geotectonic or hydrogeological models, there is, no real hydrogeological relevance to the myriad photolineaments portrayed on the Mylar satellite image overlays. Moreover, the usefulness of fracture-trace techniques diminishes dramatically as the complexity of physiography increases, and northwest. Somalia possesses a highly complex terrain. The fractured rock water-resources mapping techniques applied during BCI's study in Somalia were developed through many years of testing in New England's glaciated, culturally masked, heavily weathered, and vegetated crystalline rock terrains to identify optimum locations for high-yield wells within the full hydrogeological context of complex bedrock aquifer systems.

Remote Sensing Data Acquisition and Interpretation   This was a pioneering remote-sensing project incorporating LandSat MSS imagery and high-resolution contact prints from the ITEC-built large format camera (LFC), carried aboard NASA's Space Shuttle Challenger in October 1984. NASA supplied stereo-paired photographs of the extreme westernmost part of the Study Area, and the BCI Team used the extraordinary 60,000-km^2 "footprint," 5-m resolution, and high planametric accuracy of enlarged LFC prints both as "ground truth" for BCI's geological interpretations of LandSat MSS image maps, and as road maps, navigating for two years through thousands of square miles of poorly mapped, highly dangerous terrain in Somalia and later throughout Sudan's Red Sea Province on a separate USAID ground-water mapping project. The team also integrated specially processed satellite MSS image-data and British Ordinance aerial photographs into its map database for field exploration.

BCI's satellite image and aerial photography analysts first under-stood the geological history and structural environment of the Study Area and then focused on identifying true surface expressions of underlying geological structures consistent with the known tectonic fabric of the region, leading to realistic working maps of probable frac-ture and fault systems at a common scale with other working map prod-ucts. In the final phase of this part of the analysis, correlations were made between these newly (remote-sensing) interpreted "structural lineaments" and BCI-constructed tectonic and hydrogeological models, verified by published geological maps and field mapping. In the evalu-ation and selection process that led to the delineation of the Favorable Zones for further hydrological analysis and fieldwork, the significance attached to a structural lineaments increased with strength of expres-sion, overall length and spatial correlation with mapped faults, and frac-tures shown in the fault and brittle feature map, particularly those faults belonging to one of the three major structural trends described in the full report.

Regional Analysis and Conceptual Hydrogeological Model   An investigative hydrogeological analysis methodology was established to quantify the amount of active recharge potentially percolating through the vadose zone and available for recharge to deeper alluvial aqui-fers and conductive fracture zones feeding regional groundwater flow systems in northwest Somalia. The methodology developed synopti-cally evaluated hydrometeorological data over the Study Area, includ-ing identification of the dominant variables and parameters affecting recharge. Variables evaluated included topography, elevation, slope, evapotranspiration, surface roughness and permeabilities (i.e., caliche layer occurrence), soil infiltration rates and depth, wadi (i.e., wadi) bed characteristics (same as surface and soils), and wadi bed surface area. These variables were investigated using data from previous studies, and in the literature plus results from new fieldwork.

Although several reporting stations had documented northwest Somalia's annual precipitation, and rainfall maps of northwest Somalia had previously been published, meteorological stations were sparse, daily precipitation was not recorded, and the methodology and assumptions upon which rainfall maps were based were uncertain. The BCI team, therefore, sought a means of working with existing data to make it more useful by establishing correlative relationships between documented precipitation and Study Area topography, thereby generating a rainfall map reflecting statistical correlations of average annual rainfall with elevation. Linear regression was done with rainfall as the dependent variable and elevation as the independent variable.

Wadi beds provide an efficient pathway to the water table and, therefore, a viable mode of recharge. No information on rates of recharge through wadi beds was found. Thus, fieldwork was carried out to estimate recharge through wadi beds. Infiltration rates of the wadi bed surfaces were tested in the field to determine just how quickly storm water would be absorbed. Surface areas of the wadi beds were measured to provide relationships between infiltration rates and volume of recharge.

The slope of the land surface affects the ratio between infiltration and surface runoff. The steeper the slope, the greater the percentage of surface runoff leading to flooding conditions in the wadi beds. Steep slopes also generally have thinner soil cover and exposed bedrock, which increases the chances of groundwater recharge from the infiltration that does take place.

Field Hydrogeological Methods   Infiltration tests were performed to measure recharge to groundwater in the semiarid environment of the Study Area. The recharge potential of an area is dependent on a number of factors that are heavily influenced by the local climate such as precipitation, evaporation, vegetation, slope, ground surface condition, and stream drainage density. As rainwater falls, a portion infiltrates into the soil and the remainder is lost to evaporation or travels as overland flow. This partitioning between infiltration and overland flow is what was principally investigated in the field.

To aid in understanding the infiltration characteristics, field methods were established that allowed for many tests to be done over a large area relatively quickly. A double-ring infiltrometer was used to measure the amount of water infiltrating over time. The device, consisting of two concentric cylinders, limits the amount of lateral flow and therefore produces results primarily reflecting gravity seepage. Field procedures were conducted using two metal cylinders approximately 20.5 and 41 cm (8 and 16 in) in diameter. These devices are driven into the ground with the smaller cylinder centered inside the larger one. Water is ponded to a depth of approximately 2.5 cm (1 in). Water is added at selected intervals of time so that a constant head is maintained. These tests were run for at least one hour or until a constant rate was established.

Fifty-seven infiltration tests were performed in various surficial materials at the sites visited. The results of the tests are summarized in Table 1. Of primary importance in this table are the listings of final rates of infiltration. These values reflect the ability of the soil to absorb water under ponding conditions. It is difficult to interpret these data because the values do not reflect infiltration under natural rainstorm events. This is due to the head of water maintained in the cylinders as well as disturbance of the soil surface when driving the cylinders into place. These two factors can create anomalously high values of soil infiltration capacity. Exactly what factor to use to adjust these measured values of infiltration so that they represent true conditions is difficult to determine.

TABLE 1. Summary of Infiltrometer Tests

Test	Location	Soil/Tug	Slope	(mm/min) Infiltration Rate
1	Agabar	Soil	low	3.0
2	Agabar	Soil	low	1.5
3	Agabar	Soil	low	1.0
4	Agabar	Soil	low	1.0
5	Agabar	Tug*	low	3.0
6	Agabar	Tug*	low	5.3
7	Agabar	Tug*	low	21.0
8	Agabar	Tug*	low	45.0
9	Agabar	Tug	low	231.0
10	Agabar	Soil	low	0.9
11	Agabar	Soil	low	0.9
12	Agabar	Soil	low	0.3
13	Agabar	Soil	low	2.1
14	Borama	Soil	med	1.2
15	Borama	Soil	med	1.4
16	Borama	Soil	med	1.4
17	Borama	Soil	low	0.6
18	Borama	Soil	low	0.6
19	Borama	Soil	low	0.6
20	Borama	Soil	low	0.6

TABLE 1. *Continued*

Test	Location	Soil/Tug	Slope	(mm/min) Infiltration Rate
21	Borama	Soil	med	0.5
22	Borama	Soil	med	0.6
23	Borama	Soil	med	2.7
24	Borama	Soil	med	2.7
25	Borama	Tug	low	12.0
26	Borama	Tug	low	23.0
27	Borama	Tug	low	33.0
28	Borama	Tug	med	353.0
29	Dibrawein	Soil	low	0.6
30	Dibrawein	Soil	low	1.8
31	Dibrawein	Tug	low	3.6
32	Dibrawein	Soil	med	0.4
33	Dibrawein	Soil	med	0.8
34	Dibrawein	Soil	med	0.9
35	Dibrawein	Soil	med	0.6
36	Dibrawein	Tug	low	21.0
37	Dibrawein	Tug	low	15.0
38	Dibrawein	Soil	med	0.5
39	Dibrawein	Soil	low	0.5
40	Dibrawein	Soil	low	1.0
41	Dibrawein	Soil	med	0.9
42	Dibrawein	Soil	med	1.5
43	Jire Tug	Tug	low	5.4
44	Jire Tug	Tug*	low	4.5
45	Jire Tug	Soil	med	0.7
46	Jire Tug	Soil	med	0.8
47	Waragadighta	Soil	low	3.0
48	Waragadighta	Soil	med	1.4
49	Waragadighta	Soil	low	3.2
50	Waragadighta	Tug	low	15.0
51	Durdur River	Tug*	low	1.8
52	Biji River	Tug*	low	2.4
53	Biji	Soil	low	2.0
54	Ari Adeys	Tug	low	390.0
55	Ari Adeys	Soil	med	0.6
56	Saba'ad	Tug	low	390.0
57	Saba'ad	Tug	med	0.8

* Saturated.

Field Hydrogeological Data Analysis   The initial approach to using the infiltration data was to correlate the results with the various mapped soils so that a regional infiltration map could be generated. This approach was found to be unworkable, because existing soils classifications used by prior investigators were not appropriate for this application. For example, in Agabar, one soil type has been mapped for the area. However, it was apparent upon field inspection that many different soil types exist within even a small part of this area, varying in texture, surface condition, and depth. Such differences are directly related to slope. High-slope soils are characterized as coarse-textured and shallow in depth. Low-slope soils consist of much finer materials and are much thicker. The published Sogreah Soils Map was based primarily on local geology maps, and the soils were presumed to be the weathered product of the local rocks. Yet within these broad soil classes, natural weathering and sorting processes are controlled by rainfall and slope, often creating highly land cover exhibiting heterogeneous hydrological properties within the same soil class. This is especially true for the hydrodynamic soil properties where the flow of water and the storage of water over time are the primary considerations, not the chemical composition of the soil.

Due to these heterogeneities in local conditions, BCI established a new method that better addresses relevant soil characteristics in the target Hydrogeological Provinces of the Study Area, requiring field identification of infiltration characteristics of all major soils for the area by performing infiltration tests in each environment, thus building a new, groundwater-recharge relevant soils taxonomy for targeted Hydrogeological Provinces.

Potential recharge is defined here as the quantity of water that reaches the water table. It does not consider loss through evaporation after the recharge process. It must be stressed that potential recharge does not equal water obtainable through pumping as explained later in this section.

Quantitative estimates of recharge through the soil horizons for each basin in the Study Area were accomplished using a procedure that combines field observations with accepted hydrologic analysis techniques. Input data for each basin are summarized in Appendix A of the full report. Estimates of average annual recharge rates were generated by combining Soil Conservation Service (SCS) [McCuen, 1982] techniques for the determination of surface runoff with simple soil physics theory for the incorporation of soil moisture deficits. The SCS method uses a land classification factor CN, known as the curve number, which reflects runoff volume and moisture retention. The value of CN is determined from information on vegetative cover, land use, soil surface conditions, and soil type.

Field infiltrometer tests of soils indicated that a thin soil surface crust of caliche (generically used here to describe low-permeability soil crusts) existed in all areas visited, with the exception of wadi beds, which are addressed separately. This caliche appears to inhibit the infiltration rates outside drainage artifacts and is thus the primary factor influencing the value of CN. Because of this common occurrence, the entire Study Area was assigned the same CN of 90, representing a surface condition of scant vegetation and a soil with a low infiltration capacity. This value ignores substantial wadi bed infiltration during storm events, which is described below.

Empirical analysis was performed for the development of the SCS rainfall runoff relationship, which made it possible to estimate the amount of runoff from storm events of various magnitudes. The analysis was done on a daily basis, meaning that individual storms rather than annual or monthly averages of rainfall are used. It was assumed that only storms exceeding 10 mm (0.4 in) overcome evaporation demands, and thus have potential for groundwater recharge. The admittedly sparse data on daily storm events exceeding 10 mm for various stations in the Study Area [Sogreah, 1981A] were used to develop relationships between average annual precipitation and precipitation over the 10-mm threshold. These relationships were developed because daily rainfall data are limited to only a few weather stations. As this analysis covers the entire Study Area, known rainfall characteristics had to be extrapolated for most of the area.

Characteristics for storms exceeding 10 mm in the study were analyzed in terms of expected durations and intensities. Based on the expected storm characteristics, the value of precipitation was further modified to reflect the annual cumulative loss due to initial abstraction demands.

Recharge through the soil zone was then calculated using simple water balance theory:

$$SR = A(0.001P - Q - SM - E)6$$

where SR = soil recharge (m³/year), SM = existing soil moisture deficit, E = evapotranspiration (m/year), and A = drainage basin area (m²) (m/year).

Estimates of recharge into the wadi beds were accomplished by combining information on storm runoff characteristics, field team-measured wadi infiltration rates, and areas of the wadi beds. By combining storm hydrographs with wadi infiltration rates, the quantities of infiltration for the various basins were estimated. Wadi infiltration rates were determined by analysis of field data on saturated and unsaturated hydraulic conductivity of wadis. For each basin, the wadi beds source areas were measured. This source area was characterized as any stretch of wadi bed of sufficient width and available alluvium to allow for the acceptance and storage of large quantities of water. The total recharge for each basin is the sum of recharge through the soil and recharge through the wadi.

The technique for estimating total recharge as described above was calibrated against flood flow gauging records for the Biji Basin [Sogreah, 1981B]. Estimates of annual flood volumes and wadi infiltration derived from the methods described herein agreed with these records. Upon calibration, the method was then used to estimate recharge rates for every basin in the Study Area. These results are summarized in the full report, along with a table ranking each basin by potential recharge.

Potential recharge is defined as that quantity of water that reaches the water table. It must be stressed that upon reaching the water table, water is still available for evaporation. This is especially true in wadis where after a storm, the wadi surface is close to saturation. High rates of evaporation will occur until the water table falls below a depth of several meters. Evaporation is highly dependent on site-specific hydrogeology and varies considerably. One method of decreasing evaporation is through pumping, whereby the water table is drawn down, minimizing the quantity of moisture within reach of evaporation effects. Also, potential recharge is not the same as the quantity of water that is obtainable by pumping a well. Only a certain percentage is available, which is dependent on evaporation, well field construction, and the local hydrogeology.

The methodology presented above used assumptions that were necessary due to the lack of detailed hydrologic data for the Study Area. The primary assumptions are listed below:

- Only precipitation falling within a particular basin of interest can become recharged in that basin.
- Storms of less than 10mm are lost to evaporation. Storms over 10mm produce surface runoff and potential recharge.

- During storm events, evaporation occurs at the potential rate.
- The soil moisture deficit that exists in a given soil before a storm is equal to the wilting point of that soil.
- The soil infiltration rates are assumed to be uniform over the Study Area as supported by field infiltrometer tests.
- Slope is an adequate index of soil thickness.
- The frequency of storms over 10 mm is directly related to average annual rainfall.
- Recharge estimates neglect groundwater underflow between basins.
- Equation 4 adequately represents the time base of floods.

This study was limited by the lack of hydrologic data for the area and the absence of any type of previous detailed hydrologic monitoring program. Future work should focus on developing a data collection program designed to address the relationship among rainfall, surface runoff, and groundwater recharge. The recharge values presented in this report should be viewed as good first estimates that can be greatly refined with further work.

Detailed Geological Mapping and Geophysical Survey Methods
Fieldwork at each site began with geologic mapping to confirm the existence and assess the favorable nature of structures indicated by remote sensing and geologic maps. Such structures often coincide with straight wadi segments. Bedrock exposed in and along the banks of the wadis was examined, lithologies were noted, and the orientation and nature of structural indicators (joints and fractures) was recorded. Detailed geologic maps were made where this seemed useful.

Commonly, sediments in the wadi beds conceal bedrock; therefore, structures underlying the wadis cannot be observed directly. In such situations, geophysical methods were used to locate the hidden structures and to estimate the thickness of overlying sediments.

Four instruments were used for a total of 29,127 m of survey lines. An electromagnetic-induction ground conductivity instrument was used to measure apparent conductivity. Increases in bedrock conductivity may indicate zones of increased permeability (as well as other buried structures). Variations in the magnetic component of the electromagnetic field were measured with a passive very low-frequency-electromagnetic (VLF-EM) receiver, which uses very low-frequency radio waves broadcast by powerful transmitters at various locations around the world. In both electromagnetic methods, a primary electromagnetic field causes the flow of electric current in a buried conductor (such as a permeable geologic structure), which in turn creates a secondary field. The instruments measure the characteristics of the secondary field.

The third geophysical tool used was a magnetometer. The magnetometer measures the magnetic character of bedrock and overlying sediments. Thus, it can be useful in locating fault boundaries between different lithologies or structurally induced alteration within a single lithology. Finally, a scintillometer was used that measures gamma radiation. The instrument may signal lithologic changes, and occasionally, it may reveal permeable structures where such structures serve as passageways to the surface for radon gas produced by decay of radioactive elements at depth.

Geophysical methods produce indirect evidence for structures in the subsurface. Evaluation of geophysical evidence is difficult, and often a particular response pattern can be explained equally well by several different interpretations. The field crew was able to reduce the ambiguity of interpretations based on experience in fractured bedrock environments.

Essentially, the availability of several independent sources of information allowed the field crew to eliminate some interpretations and to concentrate on the ones that were supported by a redundancy of positive indicators. Nonetheless, it is important to remember that these geophysical techniques cannot measure secondary permeability directly, and that the interpretation of geophysical results, with regard to secondary permeability, involves a degree of uncertainty.

Hydrologic investigations were also carried out during the field trip. Infiltration tests were conducted on a representative sampling of surficial materials at most sites. Water quality samples were collected and tested at as many springs, wells, and water holes as possible. Subjective observations were made on storage capacity in wadi sediments and overbank materials.

Finally, observations were made on drill rig access, the needs of the local population, and any other factor that could influence the decision of whether to drill test wells.

Conceptual Hydrogeological Model of the Study Area

Introduction   A conceptual hydrogeological model of the Study Area was developed so that the various lithologies could be evaluated for characteristics pertinent to the occurrence of groundwater. A computerized literature search produced a number of papers that were screened for data regarding tectonic history, age, lithologic type, and chemical composition of the rock formations in northwest Somalia. Information about the tendency of each lithology to fracture (brittleness) and for the fractures to remain open (competency) is particularly important, as is the nature of the weathering products created by long-term erosion of each rock type. This information was used to first delineate and then identify the characteristics of each bedrock aquifer type.

The primary source of published geological data was the British map series produced in the 1950s to early 1960s. The maps permitted delineation of the Study Area into basic lithologic units but were not sufficiently detailed to identify lithological variations in stratigraphy that are critical to groundwater occurrence. For example, on some maps, quartz-rich psammites are not distinguished from the quartz/mica/feldspar schists. This is unfortunate because the more brittle and competent psammites represent a more favorable bedrock for aquifer development, and the quartz sand and pebbles weathering products of psammites are permeable, whereas the schists produce clays, which clog fracture systems and inhibit groundwater flow. The British maps also group all Jurassic limestones into one unit, which was not helpful in groundwater exploration. The BCI field team discovered that the lower limestone deposits are very rich in quartz sandstone and not as easily fractured as the overlying limestones and the overlying limestones are also more prone to karst development and thus are able to transmit greater volumes of groundwater.

Variances within lithologies were examined in the field, and the observations were incorporated into the lithology rating used in the matrix analysis (see below). The geology of the sites selected for detailed study was more accurately mapped in the field and evaluated contextually with all hydrogeological criteria.

The team's interpretation of locations, orientations, and geographic extent of possible bedrock fracture zones were made by use of regional geotectonic models and composite analyses of lineament, fault and brittle feature maps extracted from remote-sensing data, and various published geologic maps. The primary source of such information was the series of 1:125,000-scale quadrangle maps (1/20 × 1/20 sheets) published by the Geological Survey of the Somali Democratic Republic. Fault and joint data from these maps were digitized and subsequently reproduced at a scale of 1:250,000 so as to conform to the scale of the composite lineament overlays described above. Knowledgeable local guides familiar with the road conditions and routes to specific sites usually accompanied the field teams on field trips. Coincident with the lack of accurate base maps was an uncertainty over the correct spellings of villages, wadis, and other natural and cultural features. Spellings of place names referred to in this report are taken from the 1:125,000 British Ordnance Survey Topographic Map Series DOS 539.

Latitude/longitude ticks at half-degree intervals and control points (e.g., road intersections, drainage confluences, small, isolated outcrops, etc.), which were conspicuous in the LandSat images, were also included in the digitized map data. This facilitated the registration of lineament overlays with the fault and brittle feature map.

The suitability of bedrock water for human consumption and irrigation is dependent on the quality and quantity characteristics imparted to the source by the environment. To determine probable natural quality, the team investigated the gross rock chemistry of the various bedrock formations underlying northwest Somalia. As a result, certain geologic formations were determined to be unsuitable for ground-water development due to probable natural mineralization of the supply by the host rock. The impact of cultural activities on the quality of groundwater supplies also influenced the analysis.

Groundwater quantity from bedrock aquifers is dependent on the brittle characteristics of the rock and the hydraulic connection of bedrock fractures to recharge zones. Brittle feature occurrence, lithologic character, and potential recharge were the independently variable components of a decision matrix that was used to identify discreet hydrogeological zones where development of bedrock water appeared particularly favorable. These favorable areas were first defined on a broad scale by hydrologic province (aquifer type) and then, through further study, refined to particular geographical sites within the provinces. The drilling of high-yield bedrock wells also involved precise well-siting methods within each favorable zone because water-bearing bedrock fractures were quite narrow and torturous and masked by overburden.

Physiographic/Hydrogeological Provinces

The Inland Crystalline Highlands appeared to be the most promising for the development of previously untapped, fractured bedrock aquifers capable of providing fresh groundwater to refugee populations. The geologic structure and gross rock chemistry were favorable, and most of the refugee population in northwest Somalia was found in this area, so test drilling commenced there.

The Coastal Plain/Piedmont Province was also found to be favorable for long-term groundwater development. The oppressive heat limited permanent settlements in this area, and the extensive remote sensing and geophysical tools necessary to examine this province adequately were not available during in this project. The regional tectonic and hydrologic analyses carried out during this program did indicate that large volumes of groundwater originating in the inland provinces passes beneath the piedmont and coastal plain and is discharged as evaporation (pans are ubiquitous in the coastal plain) and directly into the sea as submarine groundwater discharge. Additional remote sensing and geophysics must be performed in these areas before groundwater balance can be estimated and favorable areas for test drilling can be recommended.

The Mid-Tertiary Plateau is subject to limited recharge, low water table, and poor water quality, but it may be possible to extract higher quality groundwater from deeper underlying bedrock fracture zones that are recharged predominantly from the crystalline province to the north. Additional work needs to be performed in this province before test drilling can be recommended.

Analysis of Principal Target Area: The Inland Crystalline Highlands
The investigation of bedrock aquifers was principally concentrated in the Inland Crystalline Highlands province, where most refugee camps were located. Eight types of aquifers were identified in this province, seven in bedrock, each identified by its predominant lithology, and the eighth in unconsolidated surficial deposits investigated here primarily for their storage and recharge relationship with the underlying bedrock aquifers. The bedrock aquifer types are as follows:

- Jurassic limestone
- Eocene limestone
- Tertiary basalt
- Cretaceous sandstone
- Precambrian granite
- Precambrian metamorphic rock
- Mafic and ultramafic crystalline rock

The Jurassic limestone has the greatest potential for the development of groundwater in large volumes, especially where fracture zones have undergone karst development.

The Eocene limestone occurs in limited areas in the northern part of the highlands. Little is known about the formation here; however, in southern Somalia and other parts of East Africa, it is a significant aquifer.

The Tertiary basalt occurs primarily as elevated, unsaturated plateaus and mesas. Where the basalt occurs in lowlands, however, it is saturated; and sand lenses, with which it is interbedded, do yield water. Vast expanses of basalt are highly permeable and may allow a great deal of precipitation and runoff to infiltrate to underlying layers.

The Cretaceous sandstone is equally favorable, although occurring only at a few locations in the crystalline highlands province. Coarse stream channel sediments within the sandstone are the most favorable water-bearing zones.

Both the Precambrian granites and metasediments have some groundwater potential; however, both the schist and gneiss components are less likely to contain open fracture systems. Both formations would have to be overlain with considerable saturated overburden before significant groundwater withdrawals could be expected.

The mafic and ultramafic crystalline rocks possess the lowest potential for significant groundwater development because their weathering products, primarily clay and calcite, tend to plug bedrock fractures.

To facilitate the collection, tabulation, and interpretation of data, the Study Area was divided into 234 fourth-order drainage basins. The numbering system for these basins was described in the full technical report, and all basins were depicted on maps accompanying the report.

Recharge of any aquifer depends on the infiltration of rainwater into the aquifer medium. The factors most influencing recharge in northwest Somalia are climatology (frequency and geography of storm events), land cover and wadi bed characteristics, and topography. Basic hydrologic data for the Study Area were sparse, and groundwater recharge calculations during this groundwater development program were based on evaluation of available local rainfall data in the context of known hydrogeological attributes of local and similar regional lithologies plus new field infiltrometer tests in representative terrains. Potential recharge estimates, representing the total volume of water reaching the water table were presented for each drainage basin in the Report, along with a hierarchical listing of basins according to potential recharge.

Two maps were generated in the full report for the explicit purpose of locating favorable areas for groundwater development from fractured bedrock aquifers. Map 1 ranked fourth-order drainage basins from A through E, with Class A basins having the greatest amount of groundwater recharge and Class E basins having the least. Map 2 ranked areas from 1 through 5 according to favorable geologic conditions for the placement of high-yield bedrock wells. Areas with Map Code I were the most favorable, and areas with Map Code 5 were the least favorable. By overlaying Mylar copies of the maps (or today using GIS technology), the areas that were favorable for both geologic and hydrologic reasons became evident.

   The result of a composite geologic and hydrologic analysis indicated that certain areas show a preponderance of favorable groundwater indicators. Some of these areas were investigated in the field using a variety of geologic and geophysical mapping methods to verify the validity of these indicators prior to recommending the drilling of test wells. Some areas were chosen for purely hydrogeological reasons, and some were chosen because there was a historical and urgent need for additional water supplies. Still others were chosen for a combination of these two factors. Seventeen sites in 12 areas were investigated using detailed field methods. The areas were identified by the name of the closest geographic feature that has a commonly recognized name. These sites were identified on a third map (see Figure 1) and include:

   (1) Agabar
   (2) Ada
   (3) Ado
   (4) *Damuk Water Hole (Borama)
   (5) *Amud Wadi (Borama)
   (6) *Dibrawein/Bira Wadi Junction (Dibrawein Valley)
   (7) *Bira Wadi (Dibrawein valley)
   (8) *Dibrawein/Soadahada Wadi Junction (Dibrawein Valley)
   (9) *Soadahada Wadi (Dibrawein Valley)
   (10) Shirwa Basati
   (11) Old Baqi
   (12) Jire Wadi
   (13) *Durdur/Waragadighta Wadi Junction
   (14) *Waragadighta Fault Junction
   (15) *Biji Valley
   (16) Ari Adeys Refugee Camp
   (17) Saba'ad Refugee Camp

Sites preceded by an asterisk, such as Borama and Dibrawein Valley, were considered most favorable for immediately accessible bedrock aquifers. Detailed geophysical surveys were conducted at two Borama sites and four Dibrawein Valley sites to precisely define where test wells should be drilled. Waragadighta and the Biji Valley ranked next

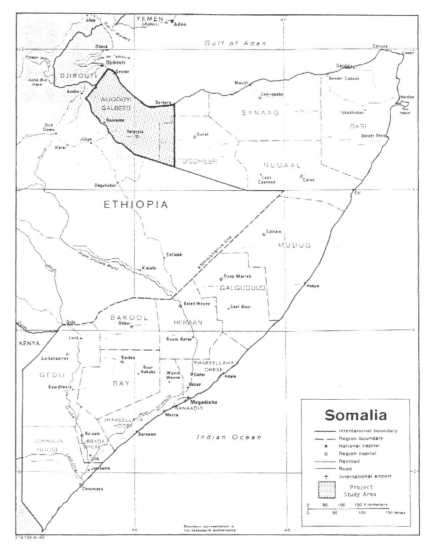

Figure 1. Map of project study area (reprinted from "Background Notes, Somalia," U.S. Dept. of State, Washington, D.C., 1981).

in favorability, with drilling pending increased settlement warranting additional water supplies.

The study consisted of both an overall evaluation of the bedrock groundwater potential of northwest Somalia and a more detailed examination of specific areas where conditions appeared quite favorable for developing new groundwater supplies from bedrock aquifers. No previous studies have addressed the possibility of extracting groundwater from bedrock fractures.

Final conclusions regarding the location, orientation, and extent of possible bedrock fracture zones were made by interpreting the composite lineament overlays in conjunction with fault and brittle feature maps extracted from various published geologic maps. The primary source of such information was the series of 1:125,000-scale quadrangle maps (1/20 × 1/20 sheets) published by the Geological Survey of the Somali Democratic Republic. Fault and joint data from these maps were digitized and subsequently reproduced at a scale of 1:250,000 so as to conform with the scale of the composite lineament overlays described above. Latitude-longitude ticks at half-degree intervals and control points (e.g., road intersections, drainage confluences, small, isolated outcrops, etc.), which were conspicuous in the LandSat images, were also included in the digitized map data. This facilitated the registration of lineament overlays with the fault and brittle feature map.

The fault and brittle feature data thus derived were augmented by similar types of information contained in three other published geologic maps: a 1:1,000,000-scale map compiled by the Somaliland Oil Exploration Company [1954], a 1:2,000,000-scale map of Ethiopia and Somalia [Merla et al., 1979], and a 1:1,000,000-scale geologic map of the Zeila Plain region [Hunt, 1942]. In each case, faults portrayed in these maps were transferred to the 1:250,000-scale fault and brittle feature map using the Bausch & Lomb Zoom Transfer Scope.

In much of Africa, the standard theory is that more than 90% (often 100%) of precipitation is returned to the atmosphere by evapotranspiration [Dunne and Leopold, 1978]. Although evapotranspiration is necessary for plant growth, it is considered a loss from the water budget in that it reduces the groundwater available for use.

Evaporation pans provide the best method of obtaining an index of potential evapotranspiration. Simple empirical formulas based on temperature, which provide an estimation of potential evapotranspiration, are approximate and tend to underestimate evapotranspiration considerably in arid areas. This is due to the fact that temperature is not an adequate index of solar radiation in arid climates.

Pan evaporation appears to average approximately 3000 mm (120 in) annually [Sogreah, 1981; Humphreys, Howard, and Sons, 1960]. The data found were insufficient to generate regional patterns of evapotranspiration. It can be confidently stated that potential evapotranspiration is much greater than rainfall so that there is always an atmospheric demand for water.

Land Cover/Soil Characteristics. The physical and chemical properties of land cover plays a major role in determining the volume of storm runoff, the rate of infiltration, and the quantities of groundwater recharge. In the Study Area, relevant land cover comprised bedrock, soils, and wadi beds, with most significant soil characteristics, including infiltration capacity, soil field capacity, and soil depth. These factors work together to govern the process of entry of water through the soil surface, storage within the soil, and transmission through the soil to the water table.

Data on field capacity of various soils found in the Study Area were located in Sogreah [1981]. Field capacity defines the proportion of water retained in the soil and the amount that drains to the water table. By combining values of field capacity with information of soil depth, it was possible to estimate the quantities of rainfall required to promote gravity drainage of water to the water table.

Infiltration capacity is defined as the rate at which the soil accepts water. No prior information on the infiltration characteristics of untilled soils was available, so field methods were developed to quantify this soil factor. An understanding of this factor allowed for the characterization of surface runoff/infiltration ratios for different precipitation intensities.

In areas where bedrock intersects the land surface or is just beneath a thin soil layer, a wide range of infiltration potential exists, from very high rates where rainfall and runoff actually percolate directly into bedrock fractures at or near ground surface, to zero infiltration over solid, nonporous bedrock.

Summary of Groundwater Potential of Study Area

The Coastal Plain/Piedmont Hydrogeological Province possesses varying thicknesses of sand and gravel found in fans and stream channels on top of eroded Quaternary pediplain. BCI's analysis indicates aquifers in this province include both alluvial and fractured crystalline bedrock. The Quaternary pediplain consists of alluvium that thickens significantly toward the coast resulting in the likelihood of significant coastal plain alluvial aquifers. A few shallow wells are located on the coastal plain, extracting fairly small volumes (1 to 2 l/s) of water for domestic purposes. Rainfall is very low [100–200 mm/yr (4–8 in/yr)] in coastal plain, and it is likely that most local precipitation merely satisfies the soil moisture deficit, and it does not contribute to the groundwater supply. The team concluded that recharge to deep, fractured crystalline bedrock hydrogeological units primarily occurs through underflow from the highlands and through significant observed infiltration of floodwater into piedmont wadi beds.

From the ancient village of Seylac (Zeila) east to Lugahaya, high salt content in the alluvium extends from the ocean to as much as 9 km (6 mi) inland. This salinity is evidence of the high rate of evaporation prevailing in the area and a probable zone of regional groundwater discharge.

The area from Lugahaya to Berbera has a much narrower and steeper coastal plain. Consequently, the alluvium may be thinner, more stream flow probably reaches the ocean, and the existence of a significant coastal plain aquifer is much less likely. Groundwaters flowing through these sediments probably discharge directly to the ocean without reaching the surface and producing high-salt-content alluvium.

It is very likely that the coastal plain and piedmont areas, in particular between Seylac (Zeila) and Lugahaya, represents a very large groundwater resource that is now being lost to evaporation and flow to the ocean. Although variable thicknesses of sand mask surface expressions of fractured bedrock, the geological model indicated sediments overlying downfaulted blocks of Jurassic limestone and Cretaceous sandstone. Initiating a local airborne remote-sensing survey using synthetic aperture radar (SAR) and thermal imaging was not possible in time of war and civil unrest and was not within the original scope of this investigation, but future application of these techniques will delineate the structural geology and identify fractured bedrock aquifers in the coastal plain and piedmont areas.

The central Inland Crystalline Highlands Hydrogeological Province consists of a sequence of igneous, metamorphic, and sedimentary rocks that range in age from Precambrian to Quaternary. These rocks form the mountainous area, including several large alluvial valleys to the south of the coastal plain. Most of BCI's exploration program focused on this area because it was where most of the refugee population was concentrated.

The primary aquifers in this area are in fractured bedrock systems, with a minor number of alluvial aquifers. The alluvial aquifers have been the only groundwater supply exploited to date. However, there appears to be a much greater potential for developing water supplies from fractured bedrock than has been realized previously. The alluvial hydrogeological province is also most likely the source for most of the recharge to the other two hydrogeological provinces.

The southernmost Mid-Tertiary Plateau Hydrogeological Province is overlain by Cretaceous Nubian sandstone and Eocene Auradu limestone. Characteristics of this province include limited local recharge, low water tables, and generally poor water quality. The primary recharge occurs from the crystalline rock complex to the north through fault and fracture systems. The water table will be closer to the surface, and water quality should improve closer to the recharge areas.

Coarse-grained zones of the Nubian sandstone is where primary permeability is the greatest and zones underlain by fractured crystalline bedrock will yield the most water. To date, limited quantities of water have been extracted from this aquifer, due primarily to the deep water table.

Groundwater Potential by Basin

The entire study region, including all three hydrogeological provinces, was divided into 234 fourth-order drainage basins. The groundwater potential of each basin was estimated by tabulating the hydrologic characteristics of each basin and calculating the total volume of water that reaches the water table (potential recharge).

The basins were ranked according to annual potential recharge and included on a large-scale Map accompanying the technical report. Examples of analytical outputs included in BCI's Report appear below.

SUMMARY OF THE AREAS SELECTED FOR DETAILED FIELD INVESTIGATION

As a result of the composite hydrogeological analysis, 17 sites in 12 favorable areas were selected for detailed field investigation. The sites were selected on the basis of lithology, structure, recharge potential, and socioeconomic factors. Example sites included in this book are identified by an asterisk.

- *· Agabar
- · Ada
- · Ado
- *· Borama Site 1—Damuk Water Hole
- *· Borama Site 2—Amud Wadi/Borama Gardens
- *· Dibrawein Valley Site 1—Dibrawein/Bira Wadi Junction
- *· Dibrawein Valley Site 2—Bira Wadi

∗• Dibrawein Valley Site 3—Dibrawein/Soadahadah Wadi Junction
∗• Dibrawein Valley Site 4—Soadahadah Wadi
• Shirwa Basati
∗• Old Baqi
• Jire Wadi
• Waragadighta Site 1—Durdur/Waragadighta Wadi Junction
• WaragadightaSite 2—Waragadighta Fault Junction
• Biji Valley
• Ari Adeys
• Saba'ad

DRAINAGE BASIN DESIGNATION SYSTEM

For the purposes of this investigation, the 35,000-km^2 Study Area was divided into 21 major drainage basins and subdivided into 234 subregional basins (Figure 1). A subregional basin is a fourth-order drainage basin, i.e., that area drained by a fourth-order stream system. Subregional basins are the basic unit of investigation having been determined as being significant in nature and manageable in size.

The basin numbering designation system was modeled after the USGS system with some modifications, which were necessary due to the nature of the Study Area. Subregional basins are numbered sequentially within each of the 21 major basins beginning at the headwaters of each basin and continuing to the discharge point of each basin. All major basins discharge north to the coastal plain sediments.

The Strahler method was used to designate stream order according to the following: first-order streams are headwaters, second-order streams result from the confluence of two or more first-order streams, third-order streams result from the confluence of two second-order systems, fourth-order streams result from the confluence of two third-order systems, and so on. Stream order designation, and thus basin order designation, only increases when stream systems of equal order join (i.e., a second-order stream joined by a first-order stream does not result in a third-order system, but rather remains a second-order system). Basin numbers followed by the letter "I" designate interstitial basins. Interstitial basins are delineated somewhat arbitrarily for convenience in calculating recharge from direct precipitation in the interstitial areas between minor basin confluences. They are numbered in sequence with fourth-order basins.

Fifth-order basins joining the mainstream channel downstream of the primary headwaters are grouped as regional basins. A hyphenated number and a regional-basin designation number identify subregional basins within each regional basin. The confluence of subregional basins at the primary headwater of a major basin was not designated as a regional basin.

The basin numbering system was slightly modified to accommodate an anomalous situation in major basin 11, which is drained by two sixth-order stream systems. Had the two sixth-order basins had their outlets in the coastal plain, each would have been designated as a separate major basin. However, because they join in the uplands, they have been considered as part of the same major basin with one subordinate to the other to facilitate numbering. One subregional basin and one interstitial basin, which fall in neither of the sixth-order basins but are part of the major basin system, have been designated 11026-0 and 11027-0I, respectively, to indicate their unique position.

RESULTS OF HYDROLOGIC ANALYSIS

Potential Recharge

Rainstorms in the Study Area are usually of short duration, lasting less than one day, but are often intense. Therefore, potential recharge was calculated by simulating individual, short-term rainfall events. Simple field capacity theory was followed to quantitatively establish some basic hydrologic relationships, where

Groundwater Recharge =
    (Precipitation $\times$ Infiltration Rate) – (Field Capacity $\times$ Soil Depth)

It was assumed that antecedent moisture conditions were at the wilting point, so that large soil moisture deficits always have to be overcome. The assumption is reasonable given the high evaporation rate in the Study Area. Information on expected daily rainfall rates and duration was found in Sogreah [1981]. As there are very sparse data on daily rainfall characteristics, it was impossible to apply this equation to the whole Study Area.

This equation did allow for the establishment of possible ranges of recharge as well as the sensitivity of certain parameters to recharge. Groundwater recharge increases as soil depth decreases because less percolating rainfall is necessary to satisfy the soil moisture deficit so that relatively more water becomes recharge. As infiltration rates were quite uniform for all soils tested, it was assumed that infiltration is a soil constant. Thus, the soil depth is the variable that is primarily responsible for spatial differences in recharge through the soil.

From direct field observations and geomorphological theory, thin soil cover prevails on the steeper slopes. Thus, recharge through the soil would be higher in these areas. However, this is minimized because as slope increases, surface runoff increases so that less rainfall is available for infiltration.

Modes of Recharge

Recharge to groundwater occurs by two major avenues, through soil and through wadi beds. Both pathways were investigated using the infiltrometer.

Soil Recharge. As the infiltration characteristics of the areas tested are quite similar, spatial difference in recharge would therefore be dependent on the ease with which soil moisture deficits can be overcome. Low soil moisture deficit is a response to low field capacity or shallow soil depth. Low slope areas are characterized as having thick soil horizons and high field capacities. It is conceivable that recharge might never occur in low slope areas except during the most extreme climatic conditions. This is a function of the low infiltration rates, relatively low quantities of rainfall, and large soil moisture deficits that need to be satisfied. The very high evaporation rates that occur daily tend to keep soil moisture deficits large so that this condition persists.

Generally, as slope increases, soil depth decreases and exposed bedrock is much more common. This leads to a condition of low moisture deficits. If fractured bedrock is exposed at the surface, or covered with a thin soil cover, significant recharge can occur because water can pass freely through fractured rock with minimal need to contribute to a moisture deficit.

Areas with much exposed or shallow lying bedrock are favorable for recharge. Rates of acceptance of percolating water and of recharge through bedrock are a function of many hydraulic factors of the fissures, such as the width and orientation of the fracture, and the degree of siltation, which impedes flow into and through the fracture. Once water enters a fracture, it can travel relatively unimpeded to the water table. As this process occurs quickly, percolated water can travel to depths below which evaporation can have an effect.

It is extremely difficult to estimate recharge from physically based data when the process is highly dependent on fractured bedrock, because the hydraulic characteristics of the rocks have to be considered as well as spatial differences in the various rock types. However, generalization can be made for areas based on observations on fracture density and degree of surface weathering. Assuming a constant loss for evaporation and absorption into the fracture system for each storm event, it is then possible to estimate relative recharge rates for various rock types.

One observable physical soil feature that is very much responsible for giving the soil a low infiltration rate is the presence of a surface crust, or caliche layer. Surface crusts were widely observed and are composed of very fine silty material intermixed with pebbles or larger rocks. The origin of this very fine-grained material is probably from the winds picking up the light particles and distributing them evenly over the land. This explains why surface crusts of similar compositions occur on steep slopes as well as low slopes. The crust becomes a hard, compact layer ranging in thickness from 1.5 to 6.5 mm (0.06 to 0.26 in). When infiltration tests are run side-by-side, one above crust and one below, the test below crust yields a higher infiltration rate in most cases. In general, soil below crust is made up of much coarser material than the crust itself.

Wadi Bed Recharge. Northwest Somalia has an organized surface drainage pattern, which promotes the second type of groundwater recharge, classified as recharge through the wadi bed. As storm water flows in a wadi, a head of water is applied to a limited area for a period of time, which is related to the storm duration and drainage basin size. Such storm flow promotes infiltration through the wadi bed and thus groundwater recharge. In upland mountainous areas, recharge from cobble-strewn stream channels penetrates the underlying rock through fractures and fissures. In low land areas where significant quantities of alluvium have been deposited in the wadi bed, rapid infiltration also generally occurs.

Infiltration tests were done on the wadi beds in each of the favorable areas investigated. These tests provided information on both saturated and unsaturated hydraulic conductivity of the wadi material. In general, infiltration rates through wadi beds under nonflood conditions were found by BCI investigators to be very high. Commonly observed well-rounded cobbles in wadi beds were also indicative of cobble and boulder migration-enhanced infiltration rates during high-energy flood runoff. In fact, ultrahigh infiltration rates during storm "flash-flood" events indicate that considerable recharge can occur during a large storm event. During a storm of considerable magnitude, wadi beds may be under several meters of water, providing hydraulic head to water infiltrating the wadi bed and spreading through underlying media. In instances where such wadis are underlain by highly permeable alluvials and fractured bedrock, recharge to deep aquifers from storm events may be significant. Personal observations of storm, even flash-flood runoff, made by the BCI team in Somalia and in similar, deeply incised drainage artifacts in the Red Sea Province of Sudan proved that 3 to 6 m of water could infiltrate similar wadi beds within 24 hours, and other investigators in Oman [Anderson et al., 1986], Chad [Ahmad, 1983], and Sardinia [Larsson, 1984] made similar observations in the Wadi environments. Several studies and construction programs have sought to enhance local aquifer recharge by dam construction on wadi-type drainage systems, at high costs and with mixed results [Haimert et al., 2002].

The long-term significance of widespread natural wadi-bed storm recharge to sustainable, deep bedrock aquifer yields in northwest. Somalia has not been addressed outside this study. However, given the widespread occurrence of wadis and the high transmissivity and storage coefficients of unconsolidated and (fractured) consolidated rocks known to occur in the region, such storm events could represent nature's most efficient contributor to deep aquifer recharge in that region.

INFILTROMETER RESULTS

The infiltrometer results range from 0.3 to 2.1 mm/min (0.01 to 0.08 in/min) for all tests on all soil types. For tests above the soil crust, the range is 0.3 to 1.5 mm/min. These ranges were well within possible errors created from disturbing the soil or from wormholes and the like. The tests conducted in wadi beds show much higher infiltration rates than for other local land cover (e.g., soils). Wadi infiltration ranged from 3 mm/min (0.12 in/min) to 353 mm/min (13.9 in/min), reflecting frequent changes in the depositional energy of wadis and the differences between unsaturated and saturated infiltration rates.

Potential Recharge

Basin characteristics used to calculate potential recharge and estimates of annual basin recharge were tabulated and published as a map in the full report. Table 2 ranks each of the 234 fourth-order basins according to potential recharge.

TABLE 2. Summary of Infiltrometer Test

Test	Location	Soil/Wadi	Slope	(mm/min) Infiltration Rate
1	Agabar	Soil	low	3.0
2	Agabar	Soil	low	1.5
3	Agabar	Soil	low	1.0
4	Agabar	Soil	low	1.0
5	Agabar	Wadi*	low	3.0
6	Agabar	Wadi*	low	5.3
7	Agabar	Wadi*	low	21.0
8	Agabar	Wadi*	low	45.0
9	Agabar	Wadi	low	231.0
10	Agabar	Soil	low	0.9
11	Agabar	Soil	low	0.9
12	Agabar	Soil	low	0.3
13	Agabar	Soil	low	2.1
14	Borama	Soil	med	1.2
15	Borama	Soil	med	1.4
16	Borama	Soil	med	1.4
17	Borama	Soil	low	0.6
18	Borama	Soil	low	0.6
19	Borama	Soil	low	0.6
20	Borama	Soil	low	0.6
21	Borama	Soil	med	0.5
22	Borama	Soil	med	0.6
23	Borama	Soil	med	2.7
24	Borama	Soil	med	2.7
25	Borama	Wadi	low	12.0
26	Borama	Wadi	low	23.0
27	Borama	Wadi	low	33.0
28	Borama	Wadi	low	353.0
29	Dibrawein	Soil	low	0.6
30	Dibrawein	Soil	low	1.8
31	Dibrawein	Wadi	low	3.6
32	Dibrawein	Soil	med	0.4
33	Dibrawein	Soil	med	0.8
34	Dibrawein	Soil	med	0.9
35	Dibrawein	Soil	med	0.6
36	Dibrawein	Wadi	low	21.0
37	Dibrawein	Wadi	low	15.0
38	Dibrawein	Soil	med	0.5
39	Dibrawein	Soil	low	0.5
40	Dibrawein	Soil	low	1.0
41	Dibrawein	Soil	med	0.9
42	Dibrawein	Soil	med	1.5

TABLE 2. *Continued*

Test	Location	Soil/Wadi	Slope	(mm/min) Infiltration Rate
43	Jire Wadi	Wadi	low	5.4
44	Jire Wadi	Wadi*	low	4.5
45	Jire Wadi	Soil	med	0.7
46	Jire Wadi	Soil	med	0.8
47	Waragadighta	Soil	low	3.0
48	Waragadighta	Soil	med	1.4
49	Waragadighta	Soil	low	3.2
50	Waragadighta	Wadi	low	15.0
51	Durdur River	Wadi*	low	1.8
52	Biji River	Wadi*	10\'1	2.4
53	Biji	Soil	low	2.0
54	Ari Adeys	Wadi	low	390.0
55	Ari Adeys	Soil	med	0.6
56	Saba'ad	Wadi	low	390.0
57	Saba'ad	Wadi	med	0.8

SHALLOW GROUNDWATER POTENTIAL

Although quantifying alluvial aquifers was not within the scope of this study, previously undetected shallow groundwater potential was identified by the BCI team as a result of the analysis of recharge to fractured bedrock aquifers, and alluvial areas exhibiting the greatest potential recharge were delineated on a map accompanying the full report. The greatest potential for undeveloped surficial aquifers is located in gently sloping, coarse-grained alluvial deposits overlying deep alluvium located in the lower parts of those basins receiving active recharge from extensive upland surface and subsurface drainage.

Surface and Groundwater Quality

Surface and groundwater qualities were measured at sites where geophysical exploration was conducted. The interpretation of these new data and water quality data from Sogreah [1983] formed the basis of this dialogue. The primary water quality problems in the Study Area were bacterial contamination and high total dissolved solids (TDS), primarily sodium, chloride, and sulfate.

Bacterial contamination was caused by human and animal waste entering surface waters and shallow wells. There were virtually no surface or groundwater protection measures in place near existing water sources. The BCI team provided local engineers and water system managers with assistance in correcting some of the old problems while drilling new, deep bedrock wells with regional recharge and little hydraulic connectivity to local surface water and resulting resistance to the extraordinary concentration of contaminates commonly surrounding desert water sources.

Excessive TDS was due to the concentrating of dissolved solids by high evaporation rates and the leaching of salts from Cretaceous and younger marine sedimentary rocks. TDS increases as surface waters and shallow groundwaters flow from the highlands to the lowlands. High rainfall and lower evaporation rates in the mountains results in higher quality water (including shallow and deep aquifer recharge) than the coastal plain, which receives little rain and has a much higher evaporation rate. Groundwater in fractured crystalline bedrock flows from recharge in the highlands to the lowlands and the sea through deep pathways uninfluenced by evaporation or human and animal contamination, and deep bedrock wells drilled by the BCI team exhibited higher water quality than local surface or surficial sources. The team surmised that accurately located deep wells drilled in the coastal plain could intercept substantial sources of mountain-recharged, high-quality water in aerially extensive confined aquifers.

The Cretaceous sandstones were submerged beneath the sea during the deposition of the younger limestones, and seawater was trapped in the in the crystal lattices of the limestone, also displacing fresh water previously trapped in the underlying sandstones. The weathering of these rocks causes increased TDS of local surface and groundwaters in many areas. However, the report indicates high-quality groundwater potential in areas of heavy fracturing, where such sedimentary units show evidence of enhanced permeability and active groundwater flows over recent geologic time, leaching out ancient salts and making those areas favorable for groundwater exploration.

Minor water quality considerations (considering alternative sources) in the Study Area include elevated hardness from dissolution of limestones and elevated iron concentrations in waters flowing through mafic and ultra mafic crystalline rocks.

Aquifer Types According to Lithology

The three main hydrogeological provinces can be divided into aquifer types according to lithology. The inland crystalline province has the greatest diversity of aquifer types, whereas both the coastal plain and the Mid-Tertiary plateau provinces are considered to be primarily monoaquifer systems for the purposes of this study. The latter two provinces could also be subdivided into discrete aquifer types; however, more detailed subsurface geologic data would have to be collected. The aquifer types would be among the same types found in the inland crystalline province. The important characteristics identified for each aquifer type are lithology, tendency to form open fractures, and recharge capability. The determination of aquifer types was based on interpretation of published data, remote sensing, and field observations.

Aquifer types in the inland crystalline province are described below.

Jurassic Limestone Aquifers

The Jurassic limestone has the greatest potential for groundwater development of any rock type. Observations made in the Study Area indicate that this rock should be a significant aquifer where it is fractured or has undergone extensive karst development. The most extensive karst development is in the southwest of the Study Area near Borama where the limestone has probably been exposed continuously since the Jurassic. Karst development decreases toward the north probably because there was no hiatus between the Jurassic and the Cretaceous and because the Nubian sandstone deposited in this area has protected the limestone from karst development.

This limestone is more easily recharged in areas where the bedding is not horizontal. The greater the dip of the sediments, the greater the recharge capability. The lower section, which usually contains lenses of sandstones and conglomerates up to 90 m (300 ft) thick, may be less favorable. Neither the published geologic maps nor the maps in this report, however, discriminate between these different lithologies. In some areas, the limestone has weathered to a clay-rich residuum that reduces the recharge capability of the aquifer. This residuum tends to accumulate in some of the alluvial valleys reducing the ability of a well in the bedrock to make use of the recharge from the overlying wadi beds.

Eocene Limestone

The most recent marine transgression occurred in Eocene time and resulted in a biogenic limestone, which has been named the Auradu formation. This limestone is a very important aquifer in southern Somalia and other parts of East Africa and Saudi Arabia. However, in northwest Somalia, it is only found as downfaulted blocks near the coastal plain/crystalline highlands boundary and as cap rock atop Cretaceous sandstone in the Mid-Tertiary Plateau province.

The Eocene limestone has not been investigated in detail. No data are available on the quality of the groundwater in the formation, although there is increased salinity where the limestone is associated with the Cretaceous sandstones on the plateau.

Where saturated, the Eocene limestone may be a very good aquifer; however, drilling must be conducted to collect water quality data.

Tertiary Basalt Aquifers

The Tertiary basalts occur primarily as elevated plateaus and are generally unsaturated. In a few areas, such as Agabar and Las Dhure, they are found in the lowlands and appear saturated. Wells drilled in these areas have intersected water-bearing zones composed of sand lenses and weathered basalt. These wells are shallow, however, and no exploration has been conducted to evaluate the water potential in the entire thickness of these basalts. The possibility exists that thicker zones of alluvium might exist at greater depths, representing a significant groundwater resource.

Field observations made in stream channels indicate that there is a paucity of vertical fractures in the basalt flows as might be expected. These basalts are extremely young and have been subject to less tectonic disturbance than neighboring formations. In certain areas, however, vertical fractures resulting from the cooling of the basalts might be expected. These areas probably represent primary recharge zones. This is perhaps supported by observations made where the basalts are unsaturated and extensive, and where there is an extreme paucity of vegetation, no soil development, and a very rubbly, gravelly appearance. This apparently indicates that these flows are extremely permeable to recharge from precipitation. If this is the case, then these basalts may permit a considerable amount of recharge to the underlying basalt and alluvial layers, which would then represent a significant groundwater resource. Evaluation of this potential will require detailed geologic mapping and the drilling of deep test holes.

Cretaceous Sandstone Aquifers

In the inland hydrogeological province, the Cretaceous sandstones are found in only a few localities, primarily where there has been block faulting, which has resulted in a downdropping of both the Jurassic limestone and the Cretaceous sandstones.

These sandstones are fine-to-medium grained and vary between consolidated and unconsolidated. The sandstone in this part of Somalia may well have been terrestrial or, at least, was deposited in a near shore environment. Therefore, there may be zones within the sandstone of very coarse stream channel sediments, which would make very productive aquifers.

Precambrian Granite Aquifers

The Precambrian granites in the Study Area are productive aquifers only where they have been fractured and faulted by tectonic activity. Remote sensing interpretation, geologic mapping, and field observation indicate that the most productive fracture zones are probably related to the recent rifting in the Red Sea and the Gulf of Aden. These granites are fairly equigranular and tend to form open fracture systems as the result of tectonic stress. The schists and gneisses, however, tend to have tectonic stress relieved by movement along crystal boundaries and foliation planes and are less likely to produce open water-bearing fracture systems. The granites do not have a significant amount of porosity or secondary permeability, except where there are well-developed fracture systems.

A weathered, granitic residuum can act as a significant reservoir for the storage of groundwater. In the Study Area, however, it appears that the residuum has been largely removed by erosion resulting from rapid uplift of the Precambrian basement.

Precambrian Metamorphic Rock Aquifers

The Precambrian metamorphic rocks in northwest Somalia are quartzites, micaschists, gneisses, metamorphic volcanics, and metapelites. Tectonic stresses tend to be relieved by movement along foliation planes, which are developed to varying degrees in the rocks. Therefore, large water-bearing open fracture systems are not formed except in places where there has been very significant tectonic movement and a tremendous amount of stress. As with granitic rocks, large groundwater supplies cannot be developed unless there is an appreciable thickness of saturated permeable overburden overlying the well site.

Mafic and Ultramafic Crystalline Rock Aquifers

These rocks are similar to granites in that they are equigranular and tend to fracture rather than relieve stress by movement along foliation planes. However, because the minerals of which they are composed break down more readily and completely than do the mineral constituents of more siliceous rocks, the weathering of mafic rocks results in a relatively large volume of weathering product. The weathering product, consisting primarily of clay and calcite, tends to plug existing fractures and prevents them from transmitting groundwater. In general, therefore, mafic and ultramafic rocks are not considered good fractured-bedrock aquifers.

Surficial and Alluvial Aquifers

Although investigating alluvial aquifers was not within the scope of this project, the presence of permeable alluvial materials, as discussed earlier, is important in the site selection of fractured crystalline bedrock aquifers.

Part of our investigative scheme, therefore, has necessarily been to identify and delineate areas of significant alluvial material, which in the inland crystalline province generally occur in wide wadis or fairly large alluvial valleys.

Also, in many cases, areas with extensive fracturing have eroded to a greater extent than surrounding areas, thereby forming a lowland with significant deposits of sand and gravel. These alluvial aquifers have been evaluated more for their storage and recharge capabilities for underlying bedrock aquifers than for direct groundwater extraction.

Geologic Structure

Neotectonics and Plate Geometry   From a plate-tectonics perspective, the greater portion of the groundwater exploration Study Area is proximal to the ridge-ridge-ridge triple junction defined by the boundaries between the Nubian (or African), Arabian, and Somali plates [McKenzie et al., 1970]. Each of these three plate boundaries is an accreting plate margin; all are active seismotectonic belts that were initiated in the middle Eocene to early Miocene, about 20 to 50 million years ago [Black et al., 1972]. Two of the accreting margins occur as mid-ocean rifts and are located in medial positions in the Red Sea and Gulf of Aden. They represent the incipient stages of ocean basin development in the Wilson cycle. The third plate boundary occurs as an intracontinental rift, the East African rift system. This feature may represent a more embryonic phase in ocean basin evolution or an aulocogen (an aborted rift structure).

The Arabian plate is situated between the Red Sea rift and the Gulf of Aden rift; the Somali plate, between the Gulf of Aden rift and the East African rift; and the Nubian plate between the East African rift and the Red Sea rift.

The present position of the intersection of the three plate boundaries is in the Gulf of Aden, a little over 100 km north of the Study Area (12°06′16″ N latitude, 44°23′38″ E longitude) [Girdler and Styles, 1978]. In each case, the plates on opposing sides of the rifts are gradually moving away from each other as the intervening voids are filled with newly generated lithospheric material. In terms of absolute plate motion (relative to a fixed frame of reference), the Arabian plate is moving northeastward toward Turkey and western Iran while the Somali plate is moving in a southerly to southeast direction. Geological evidence suggests that the Nubian plate is stationary with respect to the upper mantle [Schaefer, 1975].

Linear velocities related to plate movement can be calculated for a given location based on published information on the positions of poles of rotation for each plate pair and respective angular velocities of plate movement [e.g., Morgan, 1971]. In the vicinity of the Afar, velocities associated with relative plate motions (i.e., total spreading rates) range from about 2 cm/yr along the Red Sea and Gulf of Aden rifts (2.1 and 1.8 cm/yr, respectively) to less than 0.5 cm/yr in the Ethiopian rift valley. The latter figure is commensurate with other independently derived data on crustal spreading rates in the same region. Schaefer [1975], for example, calculated spreading rates at various locations within the Afar depression by summing the cumulative horizontal displacement (heave) of all mapped normal faults and the amount of crustal dilatation associated with fissures and dike intrusion over a measured distance. Average spreading rates can then be determined if the ages of the oldest and youngest faults and dikes are known. By this method, Schaefer calculated plate separation velocities of 0.3 to 0.4 cm/yr in the southern Afar and 0.6 cm/yr in the central Afar.

On a more detailed scale of observation, the simplistic three-plate model is complicated by the presence of at least one and possibly two microplates or continental fragments. Moreover, all of the three larger plates discussed above have suffered significant amounts of internal deformation [Mohr, 1975]. The Danakil block, or "Arrata microplate" (northwest of the Study Area), lies immediately south of the Red Sea between longitudes 40 and 44°E, extending the Danakil Alps to the

Tadjoura uplift. It is now believed that this terrain is a detached fragment of continental crust, which was originally part of the Ethiopian plateau (Nubian plate). The initiation of a rift zone in what is now the Danakil depression resulted in the separation of the Danakil block from the Nubian block about 3.5 to 5 million years ago [Barberi et al., 1975; Christiansen et al., 1975]. Recent studies [Tazieff et al., 1972] suggest that the Danakil block has also underqone a counterclockwise rotation of about 18° relative to the Nubian block as a consequence of rift development. Other similar but smaller "splinters" of the Ethiopian plateau have been mapped closer to the foot of the plateau escarpment.

The long axis of the Danakil depression is structurally controlled by en echelon normal fault scarps that trend subparallel to the Red Sea rift. The central part of the depression is characterized by block faulting, open fissures, volcanism, and extrusive mafic rocks with oceanic affinities [Tazieff et al., 1972]. This zone has, therefore, been interpreted as the continental equivalent of a mid-ocean ridge segment [Christiansen et al., 1975].

Part of the Aisha block (also called the Aisha Spur) is located in the extreme western portion of the Study Area. It is generally agreed that the Aisha block is formed from strongly attenuated, sialic, continental crust, but considerable uncertainty exists regarding the origin of this terrain and its relationship to the Ethiopian plateau [c.f. discussions by Mohr, 1972 and Christiansen et al., 1975]. Some researchers [e.g., Mohr, 1972] have considered it to be an allochthonous or detached continental fragment, which has undergone rotation with respect to the Somali plate. Others have argued against this interpretation on the basis of geophysical evidence [Makris et al., 1975] and after noting close similarities in the fracture fabrics of the Ethiopian plateau and Aisha provinces [Christiansen et al., 1975].

Black et al. [1972] consider the Aisha block to be an autochthonous inlier that has been exposed due to regional uplift and erosion. However, problems arise with this interpretation if one attempts to reconstruct the preseparation plate configuration of the Arabian and Somali plates insofar as this results in an overlap of the pre-Miocene rocks of the Arabian Peninsula and the Aisha region. It seems likely that the lack of consensus that presently exists will be resolved only after additional geological and geophysical data have been collected and analyzed.

Tectonic Model and Stratigraphy   Northwest Somalia is geologically highly complex and poorly mapped both in terms of regional stratigraphy and gross structure. Physiographically, the Study Area is bounded on the west by the Ethiopian rift valley, the Gulf of Tadjoura, and the Afar depression, and on the north by the Gulf of Aden. It is transected by the southeastern plateau escarpment, a zone of intense antithetic block faulting that separates the highly attenuated continental and oceanic crust of the Afar to the northwest from the more stable highlands of the Ethiopian plateau on the southeast.

The floor of the internal Afar is inferred to be oceanic in nature as suggested by the petrology and geochemistry of eruptive lithologies. This conclusion is supported by the observation that within this region, xenoliths of continental crustal rocks have not been found. Low values of strontium isotope ratios also argues against the possibility of continental crustal contamination of magmas [Barberi et al., 1975]. In addition, seismic refraction studies in this area reveal shallow Moho depths and unusually high Vp/vs ratios (i.e., low shear wave velocities) and indicate anomalous upper mantle conditions—specifically, elevated temperatures and partial melting [Barberi et al., 1975; Schaefer, 1975]. Very low electrical resistivity values have been similarly interpreted as an indication of high temperatures at shallow depths [Berktold et al., 1975].

The southeastern plateau escarpment represents the downwarped and downfaulted margin of the Somali plate. Within this region, bedrock is pervasively fractured and faulted. Near and south of the international border, however, the plateau area is characterized by an undisrupted Mesozoic and Cenozoic age homoclinal sedimentary succession. Crustal thicknesses here are typical of "normal" continental crust and suprabasement strata exhibit a gentle regional southeast dip.

East of the Afar, the coastal plain extends from the southeastern plateau escarpment to the Gulf of Aden, and it consists largely of a prograding sequence of unconsolidated clastic material. These terrestrially derived sediments are deposited in large coalescing alluvial fans, which form a smooth piedmont slope or bajada. Locally, the alluvial deposits are interlayered with basaltic flows and volcanogenic deposits. The veneer of surficial deposits makes the bedrock geology of the coastal plain region inaccessible for direct observation and therefore somewhat problematical. Identification of sea floor magnetic anomalies in the Gulf of Aden and tentative isochron correlations have led some researchers [Girdler and Styles, 1978] to suggest that much of the coastal plain west of 45°E is underlain by 20- to 30-million-year-old oceanic crust. Accordingly, the subsurface transition from continental to oceanic crust occurs somewhere between the northern fringe of the plateau escarpment and the coast.

The tectonic style of the greater portion of the Study Area is characterized by block faulted mountain ranges and intermontane valleys bearing a close resemblance to the Basin and Range province of the western United States [Black et al., 1972]. The trends of major fault belts within northwestern Somalia can best be understood in terms of extensional tectonic stresses acting at right angles to each of the three plate boundaries described in the preceding section of this report [Mohr, 1975]. Accordingly, various authors [Black et al., 1972; Schaefer, 1975; Kronberg et al., 1975; Azzarolli and Fois, 1964; Tazieff et al., 1972; Somaliland Oil Exploration Co., 1954] have described three predominant fault trends in this region, each being parallel to and tectonically associated with one of the major rift zones. These fault systems comprise mostly normal faults, which have formed in response to widespread crustal attenuation and dilatation. Other genetically related tectonic features included in each trend are dikes, open fissures, eruption fissures, and joints. The three main trends that are widely described in the literature are the Red Sea trend, NNW-SSE to NW-SE; the East African Rift trend, SSW-NNE to SW-NE; and the Gulf of Aden or Sheba Ridge trend, WNW-ESE to WSW-ENE.

Fields of strongly tilted fault blocks are the most striking geomorphic features in this part of Somalia [Black et al., 1972]. The majority of faults dip steeply, in the range of 700 to 800 [Somaliland Oil Exploration Co., 1954]. Bedding dips in downthrown fault blocks are typically in the range of 150 to 400 although dips as great as 600 and even 900 have been observed in some places [Black et al., 1972; Morton and Black, 1975]. Bedding dip directions are approximately perpendicular to the strike of master faults which are both antithetic and synthetic with respect to large scale crustal flexures [Black et al., 1972].

Occasionally, minor drag and drape folds are associated with faulting [Somaliland Oil Exploration Co., 1954]. Early reconnaissance mapping in this area attributed tilted strata to folding and diastrophism associated with previous orogenic events. However, more recent, detailed work has revealed that block faulting is the cause. It is believed that faulting first occurred in the Late Oligocene to Early Miocene, and it has continued episodically thereafter through Holocene time [Christiansen et al., 1975].

Field observations have shown that on a regional scale, the age relationships among the three major fault systems are both unclear and inconsistent, thus leading some researchers [e.g., Christiansen et al., 1975; Black et al., 1975; Somaliland Oil Exploration Co., 1954] to conclude that they were approximately contemporaneous. But occasionally, within established fracture domains, definitive age relationships have been noted, based on consistent truncations and offsets of one fault trend by another. In some cases, such observations have enabled workers to decipher a tentative chronology for fault development in specific areas, as in the northern Aisha domain as discussed by Black et al. [1975].

Nonetheless, it is generally agreed that the interaction of stress fields associated with the Red Sea, Gulf of Aden, and East African rifts has been both complex and time variant. Because the nature of these interactions ultimately determines crustal stress trajectories and orientations of resultant faults, it is not surprising that most fault and brittle feature maps of northwestern Somalia show a bewildering array of curvilinear fault swarms and fractures.

It is also well known that the formation of faults causes drastic perturbations of preexisting stress trajectories in areas adjacent to the faults [Chinnery, 1966]. This partially explains the presence of many aberrant subsidiary or ancillary faults that do not fall into one of the three major tectonic trends. Other evidence indicates that reactivation of old faults and other preexisting zones of crustal weakness may also be of fundamental importance. For instance, it has been suggested that foliation trends in crystalline basement rock have exerted a controlling influence on the orientation of Neogene faults [Somaliland Oil Exploration Co., 1954]. Therefore, tectonic heredity may locally be an important factor in explaining neotectonic fault patterns.

Our research has suggested that important secondary fault trends may be developed as steeply dipping transcurrent faults that are related to crustal dilatation along the three primary normal fault trends. These transverse structures would accommodate differential movement along en echelon normal fault segments or grabens and would be analogous to the Garlock fault in California as interpreted by Davis and Burchfiel [1973]. This type of fault would strike at high angles to normal faults and grabens and terminate at its intersection with these features, thereby "connecting" offset rift segments, much the same as oceanic transform faults. Although the existence of such faults in the Study Area is not widely acknowledged, it is of interest to note that similar structures have been mapped in the Basin and Range [Brun and Choukroune, 1983] and have been inferred to exist in the Afar Depression [see Figs. A and B in Barberi and Varet, 1975]. More importantly, Hunt [1942, p. 199] has mapped several northeast trending faults in the Zeila Plain of Somalia that, in this context, may be interpreted as transverse structures related to normal faults of the Red Sea trend. Accordingly, the exploration model developed for fractured bedrock aquifers regards this trend as potentially favorable.

The subsurface geometry of normal faults in northwest Somalia is largely enigmatic due to a dearth of deep drill hole and seismic reflection data. Two basic types of models have been proposed for crustal attenuation with block faulting and tilting of upper crustal layers [Black et al., 1972; Morton and Black, 1975]. Both models have specific implications regarding the geometry of normal faults in cross section. The first hypothesis [Black et al., 1972] requires that normal faults are listric faults; that is, their dip gradually decreases with depth such that they attain a "sled runner" appearance in cross section. As a consequence of listric subsurface geometry, increasing amounts of net slip on normal faults produces an increase in bedding plane dip in the downthrown block. Listric geometry may be the result of the Griffith-Coulomb failure criteria and increasing confining pressure with depth, or it may be caused by mechanical anisotropy and subhorizontal compositional layering in the lower crust. Upon assuming a gently dipping to subhorizontal configuration, the fault planes merge indistinctly with the brittle-ductile transition zone such that there is a progressive vertical transition between discontinuous (brittle) and continuous (plastic) deformational mechanisms [Brun and Choukroune, 1983]. Alternatively, as has been documented in some regions [e.g., the Basin and Range; Brun and Choukroune, 1983], listric normal faults flatten at depth and merge with a discrete regional decollement surface such as the basement cover interface or ancient overthrust zones related to earlier orogenic events.

The second model of crustal attenuation [Black et al., 1972] is based on the premise that a given fault maintains a nearly constant dip throughout the crustal profile, from the ground surface to the brittle-ductile transition zone. Thus, the fault plane is a planar rather than a cylindrical surface (the latter condition being characteristic of listric faults). Deep crustal creep resulting from ductile attenuation is invoked to explain a gradual but uniform decrease in fault plane dip. In simplified terms, the rotation of fault planes occurs as a consequence of viscous drag along the brittle-ductile boundary. Flow associated with plastic creep would be directed toward rift zones from plateau regions [Bott, 1971] and would be oriented subhorizontally and in the direction of dip of synthetic faults. Because of creep-induced drag at mid-crustal levels, fault blocks are progressively tilted sideways, much like a row of books on a shelf, so that initially steep faults become shallower with time. This mechanism offers a plausible explanation for the tilting of fault blocks observed at the Earth's surface.

The second model has been subsequently revised [Morton and Black, 1975] to include the formation of second- and third-generation normal faults. These faults form after the first-generation faults have been rotated to a dip of 40° or less. In this orientation, the first-generation faults are no longer favorable for continued dip slip movement under existing stress conditions (i.e., horizontal axis). Consequently, new, steeply dipping normal faults form from preexisting vertical joint planes that strike parallel to first-generation faults and are oriented nearly perpendicular to extensional stresses. They truncate older faults and, over time, are themselves rotated to gentler inclinations and the process is repeated. With this mechanism, it is theoretically possible to achieve bedding dips of 60° in fault-bounded blocks with about 50% crustal thinning. These figures are not unrealistic for the Afar and the southeastern plateau in northwest Somalia.

RESULTS OF COMPOSITE HYDROGEOLOGICAL ANALYSIS

Matrix Analysis

All data pertinent to the occurrence of groundwater were analyzed independently and collectively to determine which hydrogeological factors were of greatest significance and which areas exhibited the greatest density of these favorable factors.

A matrix analysis was performed to identify the various combinations of geologic characteristics that could occur and the relative degree to which each combination is favorable for the extraction of groundwater. This analysis was performed independently of a determination of the long-term available recharge to any given site.

An analysis was also performed to determine potential recharge in each basin (Figures 2–4). This analysis, however, estimated the total volume of groundwater available at the discharge of a basin, but not the amount of groundwater available at any other point within a basin. The geologic matrix analysis was further refined by determining which sites were within basins with a high potential recharge and their relative positions within those basins.

Geologic Matrix

The geologic matrix analysis considered three factors as most important in determining whether groundwater could be extracted from a given site: lithology, geologic structure, and surficial geology. The data

APPENDIX A

BASIN CHARACTERISTICS

BASIN #	PRECIP (mm)	AREA (km^2)	SLOPE (%)	ALLUVIUM (km^2)	LONG AXIS (km)
4105-3-I	500	344	2.1	.8	25
4006-I	450	114	4	2.3	14
4007	450	34	1.4	0	11
4008	450	64	3.6	0	14
4009	400	45	5	0	12
4010	400	61	1.8	0	16
4011-I	400	76	1.7	2.9	16
4012	350	49	3.9	0	11
4013	400	31	3.4	0	12
4014-I	350	158	3.3	3.7	17
4215-1	300	29	6.1	0	11
4216-2	300	54	5.6	0	12
4217-3-I	250	63	1.4	0	14
4018-I	300	112	2.8	4.7	17
4019	300	61	3.3	0	15
4320-1	200	125	1.8	0	19
4321-2	200	64	1.9	0	14
4322-3	200	102	1.6	0	20
4323-4-I	200	71	1.1	0	18
4024-I	200	146	1.3	10	36
5001	250	34	5.4	0	12
5002	250	26	5.4	0	12
5003	250	70	7.6	0	21
5004-I	250	53	2.5	0	21
6001	450	56	7.9	0	12

Figure 2. Basin characteristics.

APPENDIX B

BASIN RECHARGE ESTIMATES (ANNUAL)

BASIN	AREA (km^2)	PRECIP (mm)	RUNOFF (m^3)	SOIL RECHARGE (m^3)	TUG RECHARGE (m^3)	POTENTIAL RECHARGE (m^3)
1001	108	150	2898000	0	59000	59000
1002	88	150	2361000	1000	21000	22000
2001	47	200	1942000	0	18000	18000
2002	66	200	2728000	3000	24000	27000
2003	40	200	1653000	0	15000	15000
2004	25	200	1033000	0	10000	10000
2005	60	200	2480000	1000	31000	32000
2006-I	51	200	2108000	1000	22000	23000
2107-1	77	200	3182000	3000	30000	33000
2108-2	165	200	6820000	5000	57000	62000
2109-3-I	17	200	702000	0	8000	8000
2010-I	102	200	4216000	0	44000	44000
2011	31	150	831000	0	4000	4000
2012	94	150	2522000	0	40000	40000
2013-I	55	200	2273000	0	47000	47000
3001	558	350	4.8985E+07	0	522000	522000
3002	212	300	1.5266E+07	0	226000	226000
3003	42	300	3024000	3000	31000	34000
3004	236	250	1.333E+07	0	145000	145000
3005	48	200	1984000	0	53000	53000
3006-I	392	200	1.6204E+07	0	310000	310000
4001	93	500	1.2642E+07	8000	52000	60000
4002	212	500	2.8818E+07	8000	2403000	2411000
4103-1	42	500	5709000	2000	41000	43000
4104-2	32	500	4349000	6000	17000	23000

Figure 3. Basin recharge estimates (annual).

APPENDIX C

RANKING OF BASINS IN ORDER OF POTENTIAL RECHARGE

RANK	BASIN	POTENTIAL RECHARGE (m^3)
1	17002	5.8696E+07
2	17010-I	3.494E+07
3	17005	2.3354E+07
4	13001	2.0111E+07
5	11016-I	1.5595E+07
6	11040	1.5283E+07
7	13315-5-I	1.2069E+07
8	11041-I	1.1384E+07
9	13418-3-I	1.1229E+07
10	13002	1.119E+07
11	17215-4	9505000
12	11047-I	8556000
13	13032-I	8521000
14	20001	8298000
15	13104-2	8119000
16	4018-I	7965000
17	4014-I	7848000
18	4011-I	7785000
19	17020-I	7494000
20	17108-2	6831000
21	14001	6395000
22	20104-2	6325000
23	13417-2	6249000
24	4024-I	5694000
25	17017-I	5164000
26	4006-I	5047000
27	21010	4496000
28	9005-I	4283000
29	17004	4086000
30	4105-3-I	4040000
31	13006-I	3761000
32	6005-I	3712000
33	9004	3622000
34	11037	3598000
35	13010-I	3318000
36	8010-I	3061000
37	13103-1	2547000
38	4002	2411000
39	17001	2379000
40	13105-3-I	2356000

A-21

Figure 4. Ranking of basins in order of potential recharge.

for each category were weighted numerically on a scale from 0 to 10, with 10 designating the most favorable condition and 0 the least.

Lithology was evaluated with regard to competency and gross rock chemistry and rated as follows:

Lithology	Numerical Weighting
Jurassic limestone	10
Eocene limestone	8
Cretaceous sandstone	8
Tertiary basalt	8
Precambrian granites	4
Precambrian metasediments	3
All others	0

Geologic structure was evaluated for faults and brittle features as discerned from remote sensing analysis or obtained from published maps. The significance of a linear structure was determined according to its strength of expression, length, and correlation with azimuths of fracturing compatible with tectonic analysis. Geologic structure was rated as follows:

Geologic structure	Numerical Weighting
Significant structure	8
Structure	5
No evident structure	0

Surficial geology was evaluated with regard to the potential for groundwater storage and recharge in the immediate vicinity of a possible bedrock well site. Due to the poor quality of existing maps of surficial geology, soils, and topography and because it appeared likely that wadi beds, alluvium-filled valleys, and alluvial fans were the predominant zones of groundwater recharge and storage (verified in the field), surficial geology was rated entirely on the basis of the presence of alluvium. Although it was recognized that the volume of alluvium in a wadi can vary independently of the width of that wadi, only wadis wider than 60 m (about 200 ft) were considered favorable. Wadis of this width were also easily mapped from aerial photographs. Surficial geology was rated as follows:

Surficial geology	Numerical Weighting
Alluvium-filled valley/Alluvial fan	7
Wadi bed wider than 60 m	4
No significant alluvium	0

Individual Mylar maps of each geologic matrix factor depicting numerical weightings were composited, and the weightings were added together to delineate zones of different groundwater potential. All areas were then ranked and given a map code according to the following:

Total Numerical Weighting	Groundwater Potential	Map Code
25	Very high	1
19–24	High 2	
13–18	Moderate	3
8–12	Moderate low	4
<8	Low 5	

Figure 5 illustrates the matrix composition, and Table 3 lists possible matrix combinations for each map code.

Hydrologic Refinement of Geologic Matrix

Each area favorable from a geologic perspective was further evaluated for favorable hydrologic characteristics. Areas within Class A and Class B basins were given preference, and a subjective evaluation was also made of the position of favorable areas within a basin. Generally, the nearer a well is located to the base of catchment or basin, the greater the available recharge. Shallow groundwater quality, however, may decline in the lower reaches of the basin due to the effects of high evaporation rates in the wadis. Groundwater from a deep bedrock well in the lower basin should be of higher quality provided that some recharge, via bedrock fractures, comes from higher in the basin.

As recharge to a well occurs through a variety of pathways, no single hydrogeological environment within a basin is as important as the overall environment of the basin.

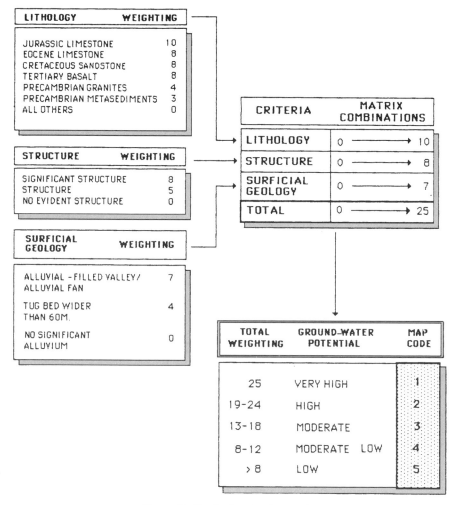

Figure 5. Geologic criteria matrix.

Favorable basin characteristics include the following:

· Uplands—significant areas of high elevation (and high rainfall)
· Midlands—high and medium slopes with little soil cover and much exposed fractured bedrock (and high infiltration)
· Lowlands—gentle topography and well-developed wadi near well site (and efficient infiltration of wadi waters and good groundwater storage capacity)

TABLE 3. Possible Matrix Combinations

Map Code	Lithology	|8|Sig. Structure	|5|Structure	|0|No Structure	|7|Alluvial Valley/Fan	|4|Tug Bed >60 m Wide	|0|No. Sig. Alluvium	Total Weighting
1	Jurrasic Limestone	✓			✓			25
2	Jurrasic Limestone		✓		✓			22
2	Jurrasic Limestone	✓				✓		22
2	Jurrasic Limestone		✓			✓		19
2	Eocene Limestone	✓			✓			23
2	Eocene Limestone	✓				✓		20
2	Eocene Limestone		✓		✓			20
2	Cretaceous Sandstone	✓			✓			23
2	Cretaceous Sandstone	✓				✓		20
2	Cretaceous Sandstone		✓		✓			20
2	Tertiary Basalt	✓			✓			23
2	Tertiary Basalt	✓				✓		20
2	Tertiary Basalt		✓		✓			20
2	Precambrian Granites	✓			✓			19
3	Jurrasic Limestone	✓					✓	18
3	Jurrasic Limestone			✓	✓			17
3	Jurrasic Limestone			✓		✓		14
3	Eocene Limestone		✓			✓		17
3	Eocene Limestone	✓					✓	16
3	Eocene Limestone			✓	✓			15
3	Cretaceous Sandstone		✓			✓		17
3	Cretaceous Sandstone	✓					✓	16
3	Cretaceous Sandstone			✓	✓			15
3	Tertiary Basalt		✓			✓		17
3	Tertiary Basalt	✓					✓	16
3	Tertiary Basalt			✓	✓			15
3	Precambrian Granites	✓				✓		16
3	Precambrian Granites		✓		✓			16
3	Precambrian Granites			✓		✓		13
3	Precambrian Metasediments	✓				✓		15
3	Precambrian Metasediments		✓		✓			15
4	Jurrasic Limestone				✓		✓	10
4	Eocene Limestone				✓		✓	8
4	Cretaceous Sandstone				✓		✓	8
4	Tertiary Basalt				✓		✓	8
4	Precambrian Granites				✓	✓		8
4	Precambrian Granites		✓				✓	9
4	Precambrian Metasediments		✓				✓	8
5	All Other Combinations							

Socioeconomic/Logistical Considerations

The selection of favorable areas for purely technical reasons was tempered somewhat by the introduction of the elements of need and logistics. Not every place in northwest Somalia with good hydrogeological characteristics is suitable for settlement or for agriculture. On the other hand, the presence of a population in need of water represents a location a high priority, whether or not the hydrogeology is particularly favorable. Finally, the very fact that a machine must drill bedrock wells requires, ultimately, that any site be accessible to the type of machinery that is practical and economically feasible to use in northwest Somalia.

The question of site access was addressed by eliminating all areas from consideration where mapped topography appeared to be too severe either for drill rig access or agriculture. Hydrogeologically favorable sites near refugee camps or near high priority areas (HPAs) were given a higher priority than sites away from such centers of obvious or potential need. Finally, several refugee camps were chosen for investigation solely because they had a critical need for additional water supplies. The socioeconomic perspective that was incorporated into the site selection process resulted from an on-going dialog with the prime contractor and Somali government officials.

Results of Matrix Analysis

All terrain that appeared reasonably accessible according to topographic maps and satellite imagery was given a designation consisting of a symbol for lithology and a numerical Map Code. Areas with Map Codes of 1 or 2 (very high or high potential) and that appear to be at good locations in basins with high potential recharge (Class A or B) were subjectively evaluated by each member of the exploration team. Each scientist ranked favorable areas, and the final selection reflected a consensus among the exploration team as to which areas were most favorable, most feasible to visit during the field trip, and would yield the most data with the geologic and geophysical survey methods that were planned. An effort was made to select a group of sites with a wide geographic distribution across the Study Area. Among the areas selected were several that were chosen solely because of an existing need for more water by the resident population.

The areas selected for detailed field study were as follows:

Agabar	Old Baqi
Ada	Jire Wadi
Ado	Waragadighta (2 sites)
Borama (2 sites)	Biji Valley
Dibrawein Valley (4 sites)	Ari Adeys
Shirwa Basati	Saba'ad

DETAILED FIELD INVESTIGATION AND RECONNAISSANCE LEVEL GEOPHYSICS

Analytical Objective

Seventeen sites in 12 areas were chosen for detailed field investigation (Figure 1). The purpose of the fieldwork at each site was to confirm the existence of the geologic structures and other site characteristics that had led to the site being classified as favorable from the matrix analysis. In addition, an effort was made to identify any other previously unrecognized physical characteristics that could influence the potential for groundwater development.

SUMMARY OF AREAS FAVORABLE FOR TEST DRILLING

Test drilling was recommended at nine sites in four areas. More than one test well may be required at each site, therefore, the priority for drilling is:

1. Borama (Sites 1 and 2)
2. Dibrawein Valley (Sites 1, 2, 3, and 4)
3. Waragadighta (Sites 1 and 2)
4. Biji Valley

All Borama and Dibrawein Valley sites were located in highly fractured Jurassic limestone possessing increased secondary permeability characteristics. As Borama and the Darai Malan refugee camp both have an immediate need for water for human and animal consumption and for agricultural development, the area was given first priority. The

Dibrawein Valley has good potential for future agricultural development and settlement, so the area is given second priority.

Both sites at Waragadighta are favorable hydrogeologically; however, agricultural potential is less than in Borama or the Dibrawein Valley. Waragadighta is on a major trade route, however. The area is given third priority.

The Biji Valley is favorable hydrogeologically; however, the remoteness of the site will hamper settlement. The Biji Valley is an HPA, and if access infrastructure is put in place, there is potential for agricultural development. The area is given fourth priority.

Should it become necessary for reasons of site inaccessibility or potential hazard to abandon test drilling efforts at one or more of the favorable areas listed above, test drilling efforts may be redirected toward the next most favorable area.

Several other sites evaluated during the detailed investigations phase of the study also are favorable for a variety of reasons, and they may appear even more favorable once direct, subsurface data are obtained from initial test wells. In fact, all sites that are indicated as being favorable on Maps 1 and 2 but were not visited should be reevaluated after results are obtained from test drilling.

GROUNDWATER POTENTIAL OF PRIMARY WATER DEVELOPMENT TARGETS

The following 10 study results will serve as examples of 28 performed, and the 10 included were the most promising.

Detailed descriptions of all the sites surveyed during the summer 1985 are in the full report. Most of the site descriptions are accompanied by large-scale location maps based on the R.A.F. air photography. The maps show major wadis, other reference points, survey lines, selected remote sensing results, and geology. For the most part, geology is taken directly from the published maps at 1:125,000 scale. However, in a few instances, the geology shown on the location map includes modifications based on field observations. Also shown on the location maps are water quality sampling points and infiltrometer test points, average rates for soil infiltration, saturated hydraulic conductivity, and unsaturated hydraulic conductivity for each site in which these tests were performed are presented for comparative purposes.

Reference: Gebile quadrangle. R.A.F. 1952 photography series 683A/143: part 1, frame 5035. Scale 1:28,475.

EXAMPLES OF ANALYSES FOR SITES CHOSEN FOR DETAILED FIELD INVESTIGATION

Authors' Note: The site evaluations presented here demonstrate the techniques applied by BCI scientists in Somalia and were chosen for the usefulness of the data to readers currently involved in northwest Somalia groundwater development.

Agabar

Introduction   Agabar is located near the northeast corner of the Gebile quadrangle. The Agabar area was targeted for detailed investigation on the basis of several criteria. Linear features detected during remote sensing correspond well with faults shown on the published geologic map. The Marodile and Las Dhure wadis join at Agabar, and together the two wadis drain a huge area. Agabar is the site of a refugee camp, with a population that could put additional supplies of groundwater to good use. Agricultural irrigation might be possible in an alluvial valley shown by the published geologic map to extend west of Agabar.

Agabar Area Geology   The structure of greatest initial interest to the BCI team at Agabar was a fault, shown on the published geologic map and visible on the 1:29,000 scale air photography, in the canyon of Marodile Wadi 3 km (2 mi) north of the refugee camp. Investigation of this structure was abandoned when it became obvious that drill rig access to any site in the canyon would be impossible.

The second area to be examined was a segment of Marodile Wadi about 1600 m (1 mi) east of the refugee camp (Figure 6). The wadi segment is broad and linear, and it coincides with a mapped fault. Metapelite, marble, and granite are exposed along the banks of the wadi. Concerning the distribution of these lithologies, agreement between published geology and field observations was good in view of the lithologic simplification, which becomes necessary in compiling a map on such a small scale.

Evidence in support of a fault concealed beneath the sediments of Marodile is indirect and weak. Lithologic contacts cannot be projected along strike across the wadi, which suggests a structural discontinuity beneath it. However, no parallel faults or attendant structures were observed in outcrop at either side of the wadi. The strongest joint sets in the area cross the wadi at a high angle. Joints parallel to the trend of the presumed fault are poorly developed.

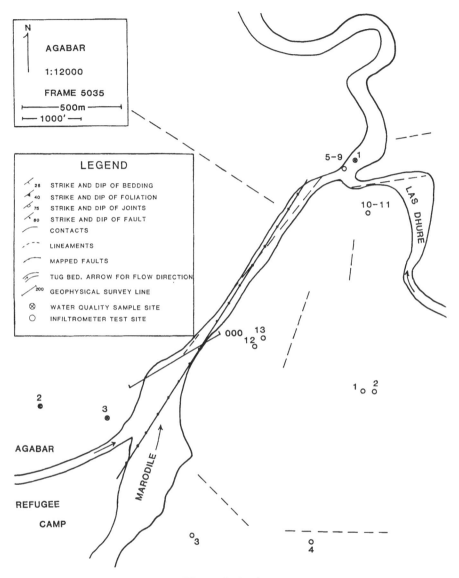

Figure 6. Agabar.

Agabar Hydrology   From a recharge perspective, this site has excellent potential. The presence of very large wadi beds, coupled with considerable exposed or thinly covered fractured bedrock, provides efficient pathways for groundwater recharge. Recharge potential at Agabar is high because precipitation is above average in the higher elevations of the large drainage basin above the site.

The data listed below (Table 4) provide a quantitative description of the recharge characteristics of the site.

TABLE 4. Agabar Site Recharge Characteristics

• Average Soil Infiltration Capacity	0.9	mm/min	(0.4 in/min)
• Average Saturated Hydraulic Conductivity of the Marodile Wadi	19	mm/min	(0.75 in/min)
• Average Unsaturated Conductivity of Wadi	230	mm/min	(9 in/min)
• Average Precipitation Over Drainage Area	425	mm/yr	(17 in/yr)
• Drainage Area	1963	km^2	($758 mi^2$)

Observations were made on the surface water flow of the wadi. The rate of flow was estimated to be approximately 9.5 l/s (150 gpm), which suggests that the wadi is at or near saturated conditions, which, in turn, indicates that water accumulates in this area. Therefore, significant water is available for use. The storage characteristics of the area are favorable. The large wadi bed volume and the surrounding expanse of alluvial fill provide considerable groundwater storage in the unconsolidated material. Alluvium is of greatest extent in the vicinity of the refugee camp. Surface condition of the exposed rock indicates that the bedrock appears to be poorly jointed and fractured and therefore has low storage potential.

The water quality of the Agabar site is adequate for a groundwater supply as indicated by the analyses of three water samples obtained near the site (Table 5).

TABLE 5. Agabar Site Groundwater Quality

No.	Source	(mumhos/cm) E. Cond	pH	(°C) Temp	(ppm $CaCO_3$) Alk	(ppm) NaCl	(ppm) Cl	(ppm $CaCO_3$) Hardness
1	Wadi	1200	7.8	30	257	500	303	428
2	Camp Well	920	7.8	30	291	350	212	342
3	Rock Well	460	7.6	27	274	400	242	223

Agabar Geophysics   Geophysical surveys were conducted in the immediate vicinity of Agabar refugee camp to verify the existence of the mapped fault/lineament aligned with the axis of the Marodile Wadi. As a result of the investigation, several small, vertically disposed, and highly resistive dike-like bodies were found beneath the wadi bed. Data obtained during mapping suggest that these NNE-trending bodies structurally control the course of the Marodile Wadi in the vicinity of Agabar. Overburden conductivities are representative of silts, sands, and coarse gravels. Overburden thickness is variable across the wadi ranging from 0 to 12 m (40 ft).

Some local overdeepening may occur in the vicinity of weathered dikes. Locally, magnetic signatures indicate the juxtaposition of Precambrian marbles against pelitic and granitic assemblages suggestive of faulting; however, the annealed character of these old fault zones would tend to preclude development of high-yield bedrock wells here.

Agabar Water Use Potential   The quantity of water available from existing bedrock and shallow sand and gravel wells reportedly is adequate for the domestic needs of the refugee camp. Currently, a limited amount of gardening is done near the wadis. Much of the alluvial area west of the camp would be suitable for farming if water was available for irrigation.

Agabar Conclusions   Combined mapping and survey results indicate that fracture systems in the area of Agabar are small and limited in lateral extent. Although old well-healed faults are apparent in area outcrops, no evidence of fault reactivation or recent fracturing was observed. Consequently, the Agabar area would not be considered suitable for the development of high-yield wells. It must be noted, however, that fractures that have direct hydraulic connections to the Marodile Wadi bed may be capable of sustained yields exceeding 3 l/s (50 gpm) due to favorable storage and recharge characteristics of area overburden materials.

Borama (Geo-Reference: Borama quadrangle. R.A.F. 1952 photography series 683A/186: part 2, frame 5023. Scale 1:29,700.)

Introduction   Borama was chosen for detailed investigation for several reasons. It is the site of the Darai Malan refugee camp, and water developed here would be useful for agriculture. Remote-sensing results for the area are positive, and the published geologic maps show faulting of Jurassic limestone against Precambrian metasediments, demonstrating that at least some structures in the area have been active since Precambrian time. The presence of limestones opens the possibility of enhanced permeability through solution enlargement of joints and fractures. Finally, springs that were noted in the area during the initial field trip provide empirical evidence of the favorable nature of the region. Two sites near Borama were investigated in detail. The city of Borama has a pumping station at a water hole on the Damuk Wadi about 3 km (2 mi) east of town. Water is also trucked from

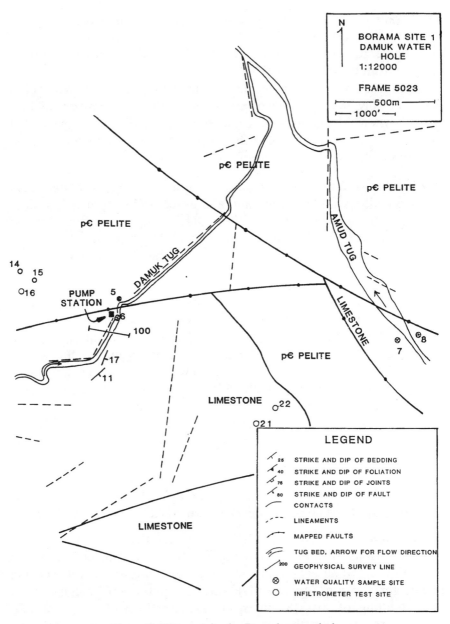

Figure 7. Borama site 1—Damuk water hole.

this site for the Darai Malan Camp. The second site is on a segment of the Amud Wadi about one mile further east. It includes a heavily cultivated area on the west bank of the wadi known as the Borama Gardens.

Borama Site 1—Damuk Water Hole

Borama Site 1 Geology   A detailed geologic map was made in the vicinity of the Damuk water hole site (Figure 8). Jurassic limestone is faulted against Precambrian metasediments about 60 m (200 ft) north of the water hole in the wadi bed, essentially as shown on the published geologic map. The limestone is poorly exposed at the water hole; but in outcrops about 50 m (150 ft) further south in the wadi, the rock is cut by fractures. The fractures occur clustered in subparallel sets, so that fracturing is focused in well-defined zones. The strike of fractures varies between 80 and 130 degrees. Fractures are vertical or dip steeply to the south.

There is abundant evidence of solution enlargement of fracture planes and selective enlargement of some bedding planes. The fracturing may be associated with the axis of an open anticline, which plunges gently to the east. The anticline is probably related to the nearby fault between the limestone and metasediments. Water may come to the surface at the water hole because it is dammed by the clay-choked fault. Two short linear segments of the wadi bed further southwest of the water hole are parallel to the local strike of the southeast-dipping limestone, and therefore are probably controlled simply by bedding. There is no evidence of fracturing parallel to these directions.

Borama Site 1 Hydrology   This site is currently used as the water supply for the city of Borama. The analysis of this site is based primarily on the shallow water well/cistern system in operation. The recharge to the area is quite high considering the limited extent of the drainage basin area, and the flow of water is remarkably constant, The fact that a well with artesian flow is located here supports the hypothesis that the limestone aquifer extends for a considerable distance away from the well location and that the recharge area extends outside the immediate drainage basin area.

The city of Borama pumps nearly 1000 m^3 per day from the water hole. As the depth of the storage tank in the wadi bed limits water hole drawdown, the system is not efficiently using the available water. A considerable increase in daily available water yield could be made through proper bedrock well location and construction techniques.

The groundwater storage capacity of the area is difficult to judge because it occurs primarily in the bedrock. Fracturing of the limestone and solution enlargement of the fractures provides storage voids. The storage characteristics of the medium could be determined through pump testing and analysis. The water quality is adequate for a drinking water supply.

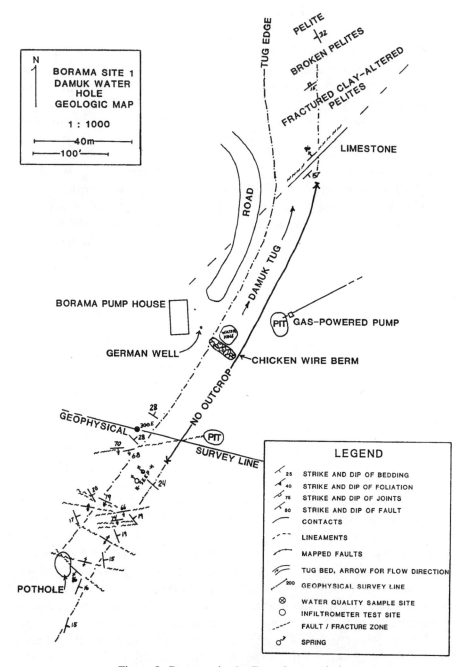

Figure 8. Borama site 1—Damuk water hole.

Borama Site 1 Geophysics   Geophysical investigations were conducted in the immediate vicinity of the Damuk water hole to ascertain the nature and extent of fracturing and faulting within the Jurassic limestone and Precambrian crystalline rocks. Results of the investigation reveal the existence of an anomalous zone that crosses the axis of the Damuk Wadi bed near the spring. The magnetic signature in the anomalous zone suggests a reduction in rock mass, possibly due to void-space development in the limestone. The radiometric signature indicates the presence of open fractures along the west wall of the Damuk Wadi bed. A local radiometric peak coincides spatially with a series of mapped fractures occurring just north of the survey line.

Borama Site 1 Water Use Potential   The Darai Malan Refugee Camp and the city of Borama would both welcome increased supplies of safe water. Local farmers have many small garden plots in the immediate vicinity of the site, some of which are irrigated using water from shallow sand and gravel wells. Additional irrigation water would substantially increase the volume of local produce.

Borama Site 1 Access   Drill rig access to Damuk water hole is good. Site preparation may be necessary.

Borama Site 1 Conclusions   The Damuk wadi site appears very promising. The site is excellent empirically in that groundwater is already being withdrawn. Hydrogeological interpretation of the occurrence is unusually straightforward, and the results are favorable for additional withdrawal. Geophysical results are compatible with the geologic observations, and groundwater quality is good. Finally, the utility of groundwater developed at this site to the local population is obvious.

Borama Site 2—Amud Wadi/Borama Gardens

Borama Site 2 Geology   The published geologic map shows that four long fault segments intersect beneath Borama Gardens, two of which are shown to separate Jurassic limestone from Precambrian metamorphic rocks (Figure 9). One of these faults is visible 1.6 km (1 mi) to the east where it crosses Damuk Wadi. Otherwise, lack of outcrop makes direct geologic evaluation of the area difficult. Overburden thickness in the gardens near the mapped fault intersection is at least 9 m (30 ft), because several wells have been dug to that depth without reaching bedrock. There is little remote-sensing evidence for the mapped

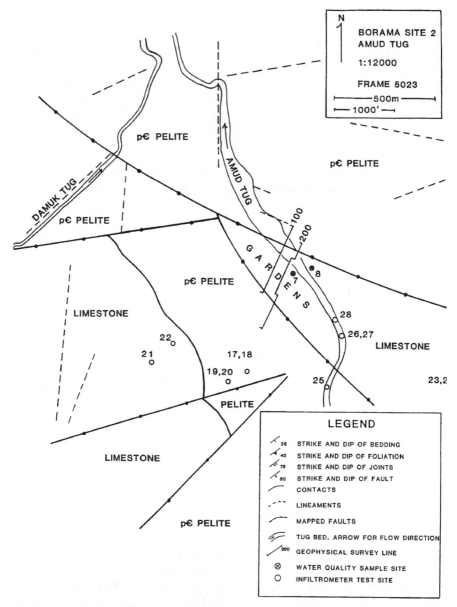

Figure 9. Borama site 2—Amud tug.

structures, but the structures would not necessarily have visible expression through the overlying sediments.

Elements of the original interpretation appear valid in spite of the lack of direct geologic evidence. Limestone crops out east of the wadi about 900 m (3000 ft) from the Gardens. By extrapolation of structure outside of the gardens, one may reasonably assume that a thickness of limestone extends into the area and that it is fault-bounded on one or more sides. Observations made at Damuk Wadi suggest that the limestone may be permeable near its faulted boundaries.

Borama Site 2 Hydrology   The Borama Gardens area has moderate recharge potential. The site is transected by a fairly large wadi, which originates in the high-rainfall, mountainous region. The only limiting characteristic is the size of the drainage basin; however, it should provide adequate recharge.

Direct water level observations were made on surrounding irrigation wells that are dug into the alluvium. These wells provide for all of the existing irrigation requirements. The information obtained through study of these wells suggests that there is considerable water available. The storage characteristics of the area are highly favorable. The wadi bed volume, coupled with a large expanse of alluvial material that borders the wadi, allows for considerable storage. It is apparent that groundwater should be available for a high yielding well. Water samples were taken directly from the local irrigation wells at the site. The quality of the groundwater is excellent.

Borama Site 2 Geophysics   Geophysical investigations were conducted near the Amud Wadi just 500 m (I 500 ft) north of the Borama road in the Borama Gardens. Here, two geophysical survey lines were laid out to detect a faulted contact between pelitic schists and the limestone unit beneath the sediments of the wadi and gardens. Although some cultural interference was encountered in the gardens, anomaly signatures can be clearly delineated amidst the noise. Results of the investigations indicate the presence of two anomalous zones located in the gardens south of the Amud Wadi. The southernmost anomaly most probably represents the fault contact between metapelites and limestones. The northern anomaly represents a fault or dike within the metapelite assemblage. Depth to bedrock varies over the traverse length but nowhere exceeds 30 m (100 ft). Bedrock in the vicinity of the anomalous zones lies between 12 m (40 ft) and 21 m (70 ft) below ground surface.

Borama Site 2 Water Use Potential   The potential of this site is similar to that of Damuk Wadi. Agriculture is well established here, and it is clear that additional water supplies would permit cultivation of more land. Local farmers welcomed the field crew enthusiastically.

Borama Site 2 Access   The flat terrain of the area should make access to the drill site easy, but land ownership in the heavily cultivated gardens could be a problem.

Borama Site 2 Conclusions   The Amud Wadi/Borama Gardens site exhibits several favorable characteristics. Although bedrock is not exposed at the site, indirect evidence and the projection of structures known to exist nearby confirm some of the premises that were used to justify selection of this site. Geophysical results appear to confirm the existence of a fault between limestone and crystalline rocks and narrow down its location. Water is manifestly present in the area, and recharge capacity is high. Therefore, a test hole drilled into fractured limestone near the fault should be successful.

Dibrawein Valley (Geo-Reference: Bawn quadrangle. R.A.F. 1952 photography series 683A/143: part 2, frames 5014 and 5015. Scale 1:32,500.)

Dibrawein Valley Introduction   The Dibrawein Valley was chosen for detailed investigation for many of the same reasons as the Borama area. Faults that were mapped in the limestones of the valley can also be recognized in air photos. Fault movement must have taken place since the deposition of the limestones during the Jurassic, and solution enlargement of fractures in the carbonate is to be anticipated. Springs are known to exist in the area, and the drainage basin position is very good. The field team evaluated four sites.

Dibrawein Valley Site 1—Junction of Dibrawein and Bira Wadis

Dibrawein Valley Site 1 Geology   This site (Figure 10) is near the south edge of an area of south-dipping limestone hogbacks separated by faults. The structural interpretation of this area by the BCI field team is similar to that presented on the published geologic map. Discrepancies between the two versions may simply reflect the inevitable imprecision that arises when complex structures are plotted at a small scale.

A fault that transects the base of a hogback dip slope northeast of the Dibrawein/Bira Wadi junction crosses the Dibrawein at an acute angle east of the junction. The trend of the fault is approximately 80 degrees. Evidence for the fault is primarily topographic, but jointing of the limestone increases toward it. Selective solution enlargement of joint and bedding planes can be seen along the edge of the hogback north of the fault. A second hogback west of the junction is slightly

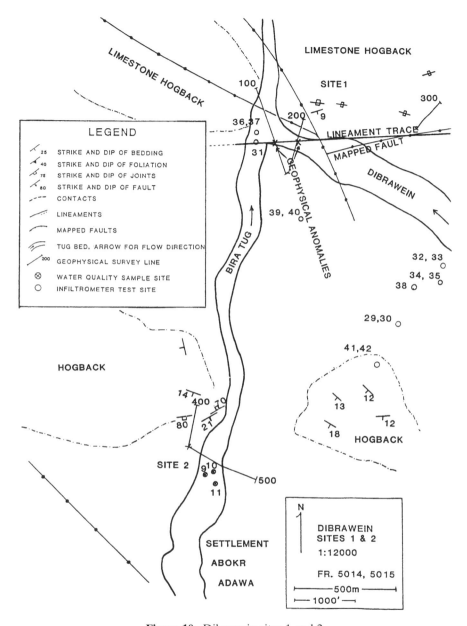

Figure 10. Dibrawein sites 1 and 2.

offset to the south with respect to its eastern neighbor. The two hogbacks may be separated by a fault underlying the north-trending segment of the Dibrawein north of the wadi junction. The straightness of the Bira Wadi south of the junction (trending at 10 degrees) and the apparent offset on opposite sides of it suggest that this wadi segment also is underlain by a fault, but there is no direct evidence from out-crops to support this idea.

Dibrawein Valley Site 1 Hydrology   The recharge characteristics of this site are highly favorable. The two wadis that join at this point drain a large area. The wadi bed material is very permeable as supported by infiltration tests. Thus, the wadi beds provide efficient pathway or groundwater recharge. The site parameters are listed below:

Average Soil Infiltration Capacity Average	0.7 mm/min (0.03 in/min)
Unsaturated Hydraulic Conductivity of the Wadi	14.2 mm/min (0.6 in/min)
Average Precipitation Over Drainage Area	475 mm/yr (<19 in/yr)
Drainage Area	1188 km^2 (459 mi^2)

One hand-dug water hole was available for inspection at this site. It appears that the water table in the wadi is close to the surface and that high water table conditions persist even during the dry season. The overall storage characteristics of the area are highly favorable. The Dibrawein Valley has substantial deposits of alluvium along the wadi beds, which provide significant long-term storage of groundwater. It is apparent that sufficient groundwater exists to recharge high-yield bedrock wells adequately. The water quality of the site should be similar to that of the samples obtained in adjacent sites in the Dibrawein Valley. The only available water point at this site is a cattle watering hole containing putrid water. Because this watering hole was polluted with animal waste and not representative of the overall water quality of the area, it was not tested.

Dibrawein Valley Site 1 Geophysics   Located near the confluence of the Dibrawein and the Bira Wadi, survey lines 100, 200, and 300 transect a topographically and tonally expressed lineament, thought to represent an intra-ormational fault in Jurassic limestones. Results of EM, conductivity, and magnetic investigations strongly suggest that faulting in limestones causes the west-trending lineament.

Dibrawein Valley Site 1 Water Use Potential   The site is about 5 km (3 mi) miles west of the new provincial capital of Baqi. The deeply weathered limestone soil that exists here should be suitable for cultivation in many areas of the broad, flat valley.

Dibrawein Valley Site 1 Access   The road through the mountains between Borama and Dibrawein is rough but passable. A drill rig should be able to get through it if driven with care. This particular drill site is easily accessible.

Dibrawein Valley Site 1 Conclusions   Geologic and geophysical evidence confirm that a west-trending fault in limestone crosses the Dibrawein near the Dibrawein/Bira junction. This fault may be intersected by other structures at the wadi junction. Observations made elsewhere suggest that the structure is permeable, and hydrologic factors are favorable. This site is considered an excellent location for a test well.

Dibrawein Site 2—Bira Wadi

Geology   About 1.6 km (1 mi) south of the Dibrawein/Bira junction, the Bira Wadi passes between two hogbacks of Jurassic limestone (Figure 10). Strikes are east to southeast. Dips range between 10 and 20 degrees south. Analogy with similar hogbacks in the area suggests that the dip-slopes of the two hogbacks flanking the Bira may be fault bounded. Indeed, a broad zone of intense fracturing is well exposed near the base of a hogback on strike with this site 3 km (2 mi) to the east. This east-trending structure may be intersected by a north-trending one coincident with the Bira Wadi. Bedrock is concealed by sediment along the Bira, but topography and structural geometry support, the existence of a fault between the two hogbacks. Specifically, the course of the Bira corresponds with a broadly defined lineament that extends for more than 5 km (3 mi), and there is an apparent offset of the two hogbacks with respect to each other.

Dibrawein Site 2 Hydrology   The recharge potential of this site is favorable but inferior to that of the other sites in this region. This is mainly a function of the smaller drainage basin that drains to this point.

Average Soil Infiltration Capacity	0.7 mm/min (0.03 in/min)
Average Unsaturated Hydraulic Conductivity of the Wadi	14.2 mm/min (0.6 in/min)
Average Precipitation Over Drainage Area	475 mm/yr (19 in/yr)
Drainage Area	251 km^2 (97 mJ:1)

The storage characteristics of this site are good. Considerable alluvium extends away from the wadi. The depth of this material is questionable because bedrock outcrops exist nearby. The several dug wells in the wadi have a water level of about 2.5 m (8 ft) from the surface. This is lower than the other sites visited in the area. Overall, this site is moderately favorable in terms of recharge. The water quality of the site is good and actually better than other sample points in the area (Table 6). This is probably a function of the low water table conditions, which minimize evaporation.

TABLE 6. Dibrawein Site 2 Water Quality

No.	Source	(mumhos/cm) E. Cond	£!!	(°C)	(ppm $CaCO_3$) Alk	(ppm) NaCl	(ppm) Cl	(ppm $CaCO_3$) Hrdness
9	Well	480	8.2	25	274	50	30.3	223
10	Well	500	8.2	26	291	38	22.7	223
11	Well	455	7.8	27	223	44	26.5	274

Dibrawein Site 2 Geophysics   Geophysical results at this site came from two survey lines located approximately at right angles to each other. The first line was intended to cross the east-trending fault near the base of the hogback west of the wadi. The second line crosses the Bira and was intended to detect the structure presumed to underlie it.

The passive EM, conductivity, and magnetometer surveys on the first line all produced evidence for a discontinuity, which probably corresponds with the postulated east-trending faulted limestone. No evidence provides the results of the second survey line to confirm the existence of a structure coincident with Bira Wadi.

Water Use Potential   The potential of this site is similar to that of the previous one. Nearby land on both sides of the wadi appears suitable for cultivation.

Dibrawein Site 2 Access   Drill rig access to the site is good.

Dibrawein Site 2 Conclusions   This site is considered favorable for the same reasons as Site 1. Geologic and geophysical evidence indicates that a fault runs along the base of the hogback west of Bira Wadi at this site, and it probably crosses the wadi. The fault may be intersected by a second structure coincident with the wadi, although there is no geophysical or direct geological evidence to confirm this. Whether the second structure exists or not, observations made at this site and elsewhere indicate that the east-trending fault should serve as an effective conduit for groundwater. Because the zone of fracturing and solution enlargement of cavities associated with the fault is probably broad, precise location of test holes here and at similar sites may not be critical.

Dibrawein Site 3—Junction of Dibrawein and Soadahada Wadis

Geology   The published geologic map shows a sliver of gabbro faulted against limestone northwest of the Dibrawien/Bira Wadi junction (Figure 11). A small wadi with a northeasterly trend, the Soadahada, joins the Dibrawein near the northwest end of the gabbro exposure. A detailed geologic map was made along the wadi segments upstream from the junction (Figure 12).

Exposures are best along the banks of the Soadahada, where the relationship between the gabbro and the limestone can be seen clearly. Fresh gabbro underlies the wadi junction. The gabbro exposed in the small wadi becomes weathered between 150 km (500 ft) and 215 km (700 ft) southwest of the junction. It is disconformably overlain by a layer of conglomerate about 20 m (70 ft) thick, which forms the base of the Jurassic sedimentary sequence. Standing water occurs in the wadi bed along most of the segment underlain by conglomerate. Limestone overlies the conglomerate. The contact is about 300 m (1000 ft) south-west of the wadi junction, and limestone exposures are nearly continuous along the small wadi from this point to the end of the survey line at 550 m (1800 ft). Bedding dips rather uniformly to the southwest at 10 to 20 degrees. Two well-defined faults were observed in the limestone along the wadi bed, striking between 70 and 90 degrees and dipping steeply to the north. It is inferred that one or both is continuous with a fault seen in the south bank of the Dibrawein about 300 m (1000 ft) southeast (upstream) of the wadi junction.

Alluvial sediments cover the bedrock in this segment of the Dibrawein, which is interpreted to be underlain by conglomerate, and the gabbro/conglomerate and conglomerate/limestone contacts are masked by the alluvium. The limestone is heavily jointed throughout its exposed extent to the end of the survey line at 335 m (1100 ft) south-east. Joint planes and some bedding planes show considerable solution enlargement in the jointed zone near the east end of the survey line, giving the rocks there a texture like Swiss cheese.

Springs occur at wadi level in limestone near the east end of the fault and in the middle of the heavily jointed zone. In the covered wadi segments, water seeps from the wadi bank near the presumed location of the gabbro/conglomerate contact. Thus, in both wadis, groundwater comes to the surface south of the contact between the gabbro and Jurassic sediments. One may conclude that the contact functions as a

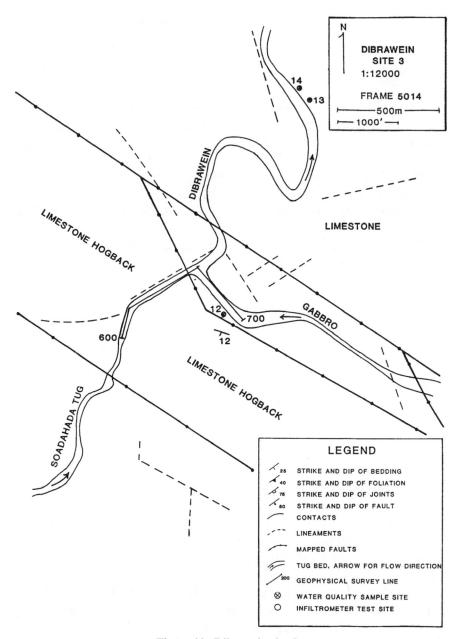

Figure 11. Dibrawein site 3.

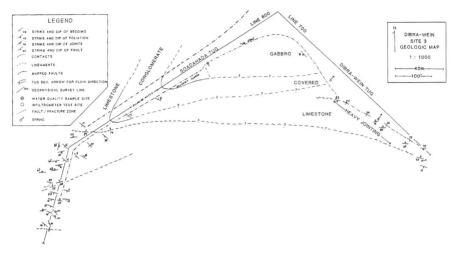

Figure 12. Dibrawein site 3.

dam. This damming effect together with the permeability induced by the fault makes the Soadahada/Dibrawein junction a favorable site.

Dibrawein Site 3 Hydrology   The recharge characteristics of this site are excellent. A very large drainage basin feeds the site.

Average Soil Infiltration Capacity	0.7 mm/min (0.03 in/min)
Average Unsaturated Hydraulic Conductivity of the Wadi	14.2 mm/min (0.6 in/min)
Average Precipitation	475 mm/yr (18.7 in/yr)
Over Drainage Area	122 km^2 (471 mi^2)
Drainage Area	0

The storage characteristics of the area are questionable. The site is surrounded by bedrock, and the storage available is difficult to determine without drilling a well. Storage would ultimately depend on the frequency and size of solution voids and fractures in the rock.

   The water quality of this site is adequate for a drinking water supply. The samples shown in Table 7 below were taken directly from the nearby springs originating in limestone.

Dibrawein Site 2 Geophysics   Survey lines 600 and 700 are located in the immediate vicinity of the confluence of the Soadahada and Dibrawein wadis. Surveys were conducted to better understand the relationship between downfaulted Jurassic limestones and faulted Precambrian basement gabbros. Results of EM and magnetic surveys

TABLE 7. Dibrawein Site 3 Water Quality

No.	Source	(mumhos/cm) E. Cond	.E!!	(°C) Temp	(ppm CaCO₃) Alk	(ppm) NaCl	(ppm) Cl	(ppm CaCO₃) Hrdness
12	Spring	950	6.7	33	308	113	68	377
13	Spring	1200	7.8	31	342	187.5	114	325
14	Spring	1150	7.8	31	342	187.5	114	325

indicate the presence of at least two faults within gabbro basement sandwiched between two relatively downthrown, limestone-capped, and gabbro-basement-cored blocks. The upthrown gabbro block extends beneath the Dibrawein Wadi. Overburden deposits in the Dibrawein Wadi bed vary in thickness between 0 and 9 m (30 ft) with some local overdeepening. The Soadahada Wadi flows directly on limestones, sandstones, and gabbros; only in the vicinity of its confluence with the Dibrawein Wadi do thin overburden deposits exist.

Water Use Potential   This site is less convenient than the two previously described. The closest established settlement is Abokr Adawa, about 4 km (2.5 mi) southeast on the Bira. However, much of the flat land south of the site along the Soadahada appears suitable for cultivation.

Access   Access to the site via the Dibrawein should be good if drilling is done at a time when the wadi bed is reasonably dry. Otherwise, a moderate amount of roadwork would be required to allow entry along the south edge of the hogback east of the site.

Conclusions   This site is very promising. A broad zone of fracturing and solution enlargement of cavities is well exposed and unmistakable. The obvious geologic interpretation is supported by geophysical results. The surface flow of water from the structure provides empirical evidence of its favorable characteristics, and the structure's intersection with the Dibrawein ensures adequate recharge.

DIBRAWEIN SITE 4—SOADAHADA WADI

Geology

The published geologic map shows that an inferred fault trending at approximately 130 degrees crosses the Soadahada about 0.8 km (0.5 mi) south of the site previously described. On 1:29,000-scale air photography, the fault is expressed as a discontinuous but definite lineament more than 5 km (3 mi) long that extends to the southeast as far as the Bira Wadi. Remote sensing also suggests that other structures following the general course of the Soadahada may cut across the fault. This is indicated by several short lineaments in the bed of the wadi and another longer feature to the west.

The northwest-trending fault corresponds with a broad zone of fractured limestone. Fault movement has not been concentrated along a few discrete planes in this zone, which is about 90 m (300 ft) wide. Instead, the rock is uniformly and generally penetrated by fractures spaced at intervals of about 8 cm (3 in) to 0.6 m (2 ft). Sufficient movement has taken place to cause substantial dislocation of adjacent limestone blocks with respect to each other, creating abundant open spaces. Bedding is difficult if not impossible to distinguish. In some outcrops, it is supplanted by rodding, evidently caused by the intersection of bedding with a high-angle fracture plane, which is repeated at roughly 0.3-m (1 ft) intervals. Other outcrops have a chaotic, jumbled appearance, like bricks dumped in a pile. Obviously, the zone is extremely permeable, and depending on the elevation of the water table at this location, the structure it represents should be large enough to serve as an effective groundwater collection system.

The northwest-trending fault is the dominant structure here, and evidence exists for movement along a structure parallel to the Soadahada. The rodding mentioned above plunges to the northwest between 10 and 20 degrees, indicating a mild flexure underlying the Soadahada. Rocks northwest of the structure are downdropped with respect to rocks to the southeast. Joints and a small fault trending approximately 40 degrees indicate a cross-cutting structure.

Dibrawein Site 4 Hydrology The site is less favorable from a hydrologic perspective than the other sites examined in the Dibrawein valley. The size of the basin drained by the Soadahada is small, and storage capacity in the sediments of the wadi bed is minimal. No surface flow was present at the site in July 1985. However, the storage capacity of the structure should be substantial, and the structure is probably large enough to be hydraulically connected with the Dibrawein drainage system.

Average Precipitation Over Drainage Area   400 mm/yr (16 in/yr)
Drainage Area   25 km^2 (10 mi^2)

Geophysics   Rock and structure are well exposed at this site, making geophysics unnecessary.

Water Use Potential   This site is not near an existing population center, however, the surrounding terrain is flat, and the soil cover is well developed.

Access: Access to the area is good, with modest road construction near the site.

Dibrawein Site 4 Conclusions   The Soadahada Wadi site appears promising for groundwater development, although somewhat removed from sources of recharge. Widespread and intense fracturing of area bedrock allows for storage of significant volumes of groundwater. Furthermore, area fracture systems are laterally extensive, thereby increasing the hydraulic connectedness of individual fractures with potential sources of recharge.

Old Baqi Site (Geo-Reference: Borama quadrangle. R.A.F. 1952 photography series 683A/143: part 2, frame 5178. Scale 1:31,740.)

Introduction   Baqi is a small settlement in a broad mountain valley on the road from Borama to the Dibrawein Valley. The field team referred to it as "old" Baqi to distinguish it from Baqi, the new regional capital in the Dibrawein Valley. The settlement is located at the junction of two wadis. The combined wadi flows north into the Dibrawein Valley, where it is known as the Bira. This area was selected for detailed investigation on the basis of remote-sensing results and observations made during the Winter 1985 field trip, when the Precambrian psammites of the area were observed to be heavily jointed and fractured.

Old Baqi Site Geology   The published geologic map shows the Baqi area to be underlain by psammites. However, the rock actually exposed near Baqi is granitoid gneiss of intermediate composition. Its igneous origin is supported by the presence of scattered mafic inclusions.

The primary structural target at Baqi is the intersection of two lineaments, coinciding generally with two of the three wadi segments at the wadi confluence noted above (Figure 13). The east-trending lineament is parallel to foliation, which dips north between 30 and 40 degrees. The development of foliation is variable, but generally it is subdued. The parallelism of the east-trending lineament with foliation cannot be seen as a favorable indication if the lineament can be explained by foliation alone, because foliation planes are not inherently

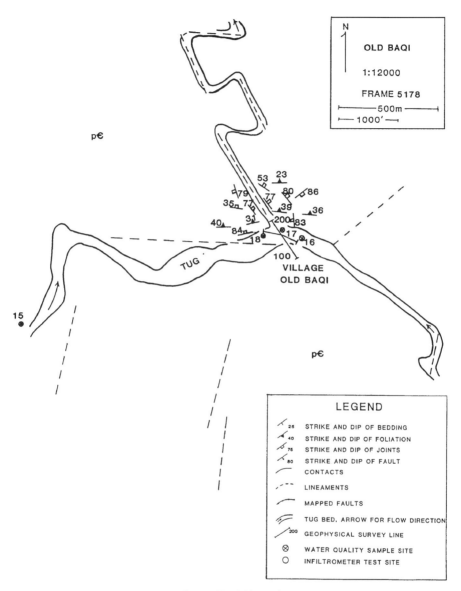

Figure 13. Old Baqi.

permeable. However, as planes of weakness, foliation planes are more likely to accommodate movement than planes at other orientations, producing increased permeability. No direct evidence can be seen for such movement near the lineament, but no outcrop was exposed coincident with it.

The second lineament trends northwest, is remarkably straight, and is parallel to the joint direction most commonly displayed in the gneiss (150 degrees). Dips range generally between 75 degrees east and 80 degrees west. Outcrop exposures are good in the bed of the wadi, and no offset can be seen there or in the wall of the wadi at the north end of the segment, where the postulated structure should be best exposed. Although rock exposed on the hillsides adjacent to the wadi appears broken and jumbled, the gneiss is fresh and joints are tight in the wadi bed where the weathered rock has been removed. This jointing, which appears to account for the lineament, could increase permeability only marginally.

Old Baqi Site Hydrology   The recharge potential of this site is moderate. The drainage basin is smaller than other sites investigated. This limits the catchment of rainfall and, as a result, recharge.

Average Precipitation Over Drainage Area	500 mm/yr (20 in/yr)
Drainage Area	166 km^2 (64 mi^2)

The storage capacity of this site is severely limited, because bedrock is very close to the surface. This fact was classically illustrated by the town's dug well. This well has a depth of about 1.2 m (4 ft) and is in close contact with the shallow bedrock. The depth of water was about 15 cm (6 in), indicating minimal storage and available drawdown. Proper well location could provide this town with a much safer, more reliable source.

The water quality of the site is excellent. The quality of the water from the two tested wells was among the best observed on the field trip. This is because evaporation through the wadi is limited by water availability and exposed surface area (Table 8).

Geophysics   Geophysical surveys were conducted near old Baqi to assess the nature and fracture character of the lineaments. Survey lines

TABLE 8. Old Baqi Site Water Quality

No.	Source	(mumhos/cm) E. Cond	E!!	(°C) Temp	(ppm CaCO$_3$) Alk	(ppm) NaCl	(ppm) Cl	(ppm CaCO$_3$) Hrdness
15	Soldier Camp	680	7.2	24	377	200	121	342
16	Town Well	470	7.7	27	171	50	30.3	223
17	Wadi	800	7.7	26.5	308	125	75.8	377
18	Wadi	1400	7.8	28	325	375	227	377

100 and 200 transect NNW- and east-trending lineaments underlying two wadi segments at the wadi junction north of the village. Results of the survey suggest the existence of a small, moderately dipping and highly resistive dike-like body located within old Baqi village. The anomaly may be associated with a west-trending lineament, but because it may also be coincident with local foliation trends, its true nature cannot be determined. Bedrock is exposed along the north banks of the wadi segments upstream from the junction. The valley occupied by Waragadighta Wadi broadens west of the fault junction. A second fault segment extending from the junction on a westerly trend is mapped as running along the south edge of the valley parallel to the break in slope between the valley wall and the adjoining north sloping pediment. This area is covered, and the field team was unable to make direct outcrop observations to confirm the existence of the fault.

The third inferred fault segment extends east from the mapped fault junction through an area of good outcrop exposure in the northeast wall of the valley. Near the wadi, it is shown to separate Precambrian psammite from pelitic schist and gneiss. No evidence for such a fault was found. Indeed, the transition between the mapped lithologies appears gradual rather than abrupt.

Old Baqi Site Hydrology   The recharge potential of the two sites is very good. The sites are near a very large wadi that drains an enormous area. Also, soil infiltration capacities are some of the highest observed in the field.

Average Soil Infiltration Capacity	2.8 mm/min (0.11 in/min)
Average Unsaturated Hydraulic Conductivity of the Wadi	15 mm/min (0.6 in/min)
Average Precipitation Over Drainage Area	425 mm/r (17 in/yr)
Drainage Area	3375 km^2 (1303 mi^2)

The wadis at both sites are very large and typical of wadis that drain a very large area. The wadis vary in width from 180 m (600 ft) to 300 m (1000 ft). Depths are estimated to be 20 m (65 ft) to 24 m (80 ft). Thus, the volume of wadi material is very large and is capable of storing large quantities of groundwater. Surface flow was commonly observed in the area, and the flow rate ranged from 60 l/s (1000 gpm) to 90 l/s (1400 gpm). These characteristics make the site highly favorable with respect to recharge.

Geophysics Survey line 300, located about 1.6 km (1 mi) south of the village of Waragadighta on the Waragadighta Wadi, was located to transect the NW-trending mapped fault and lineament within the Precambrian granitoid and metasedimetary assemblage. Interpretation of survey results, although admittedly tentative, suggests the presence of a fairly deep bedrock trough filled with resistive overburden materials ranging from sands to coarse gravels. Area bedrock resistivity data are consistent with the range of resistivity values for granitoid lithologies. Subtle variations in area magnetic profiles may be indicative of faulting within Precambrian granites/granitoids.

Water Use Potential Water developed at this site could be used for irrigation of the flat area south of the wadi. However, as a practical matter, construction of a flood-proof well and distribution system in the bed of such a large wadi would be difficult.

Access Access is good. A truck could be driven nearly anywhere on the flat, sand-covered surface of the wadi.

Old Baqi Site Conclusions Results of geologic and geophysical investigations warrant test drilling at this site to determine the fracture character and water-bearing capacity of Precambrian granites.

REFERENCES

Ahmad, M.U., 1983, A Quantitative Model to Predict a Safe Yield for Well Fields Kufra and Sarir Basin, Libya: Ground Water, Vol. 21, No. 1, pp. 58–66.

Archer, S.R. Chetwynd, 1955, Report on Bargeisa Water Supply: Government of Somaliland, Hargeisa.

Azzarolli, A., and Fois, V., 1964, Geological Outline of the Northern End of the Horn of Africa: Proceedings of the Twenty-Second International Geological Congress, New Delhi, Vol. 4, pp. 293–314.

Barberi, F., Ferrara, G., Santacroce, R., and Varet, J., 1975, Structural Evolution of the Afar Triple Junction, in Pilger, A., and Rosier, A. (Eds.): Afar Depression of Ethiopia: Proceedings of an International Symposium on the Afar Region and Related Rift problems, Vol. 1, Inter-Union Commission on Geodynamics Sci. Rep. No. 14, pp. 38–54.

Barnes, S.V., 1976, Geology & Oil Prospects of Somalia, East Africa: The American Association of Petroleum Geologists Bulletin, Vol. 60, No. 3, pp. 389–413.

Beydoun, T.R., 1970, Southern Arabia & Northern Gomalia: Comparative Geology: Phil. Trans Royal Soc. London, Vol. 267, No. 1181, pp. 267–292.

Black, R., Morton, W.R., and Rex, D.C., 1975, Block Tilting and Volcanism within the Afar in the Light of Recent K/AR Age Data, in Pilger, A., and Rosier, A. (Eds.): Afar Depression of Ethiopia: Proceedings of an International Symposium on the Afar Region and Related Rift Problems, Vol. 1, Inter-Union Commission of Geodynamics Sci. No. 14, pp. 296–300.

Morton, W.H., and Varet, J., 1972, New Data on Afar Tectonics: Nature Physical Science, Vol. 240, pp. 170–173.

Bonatti, 1976, Afar Between Continental and Oceanic Rifting: Bull. Soc. Econ. Geol., Vol. 72, No. 7, pp. 1364–1365.

Brown, C.B., 1930, The Geology of British Sornaliland: Quart. Journ. Geol. Soc., Vol. LXXXVII, pp. 259–280.

Burgett, C.L., and Heyda, C.M., 1982, Gazeteer of Somalia: Defense Mapping Agency, Washington, D.C.

China National Complete Plant Export Corp., 1983, Report on Geological Exploration of Exploratory Boreholes & Test Wells in the Northwest Region of Somalia Democratic Republic: Chinese Well-Drilling Team, Hargeisa, Somalia.

Dagg, M., and Blackie, J.R., 1970, Estimates of Evaporation in East Africa in Relation to Climatology Classification: Geographical Journal, Vol. 136, pp. 227–234.

Dunne, T., and Leopold, L.B., 1978, Water in Environmental Planning: W.H. Freeman and Co., San Francisco.

Fallace, C., 1983, A Brief Review of the Surface and Ground Water Resources of the Northwest Region of Somalia, in: Quaderni Di Geologia Della Somalia, Vol. 7, Mogadishu, pp. 171–187.

Focus on Northwest Region Agricultural Development project: 1983 Report, 2nd Annual Farmers Conference, Mogadishu.

Girdler, R.W., and Styles, P., 1978, Seafloor Spreading in the Western Gulf of Aden: Nature, Vol. 271, pp. 615–617.

Gouin, P., 1979, Earthquake History of Ethiopia-and the Horn of Africa: Bowker Pub., Int. Devel. Res. Centre, Ottowa.

Greenwood, W.R., 1976, A Preliminary Evaluation of the Non Fuel Mineral Potential of Somalia: U.S. Geological Survey Open File Report 82-788, Government Printing Office, Washington, D.C.

Griffiths, J.F., and Hemming C.F., 1965, A Rainfall Map of Eastern Africa and Southern Arabia: East African Meteorological Department, Nairobi, Vol. 1, No. 10.

Griffiths, J.F. (Ed.), 1972, The Horn of Africa, World Survey of Climatology, Vol. 10, Climates of Africa: Elsevier, Amsterdam, pp. 133–165.

Grimes, A.F., 1959, Bibliography of Climate Maps for Ethiopia, Eritrea, and the Somalilands: U.S. Department of Commerce, Washington, D.C.

Halcrow & Partners, 1982, Northwest Somalia Water Sypply Consultancy Interim Report: Woqooyi Galbeed Refugee Water Supply Programme, OXFAM, Oxford, England.

Hemming, C.F., 1966, The Vegetation of the Northern Region of the Somali Republic: Proc. Linnean Soc., London, Vol. 177, No. 2, pp. 173–250.

Humphreys, Howard and Sons, 1960, Hargeisa Water Supply Investigation: Ref. AH53, Colonial Development & Welfare Scheme D 3121, Westminster & Nairobi.

Hunt, J.A., 1939, Report No. 1 on the Geology and Water Supply of the Area Hargeysa-Haleva: Colonial Development Fund, Hargeisa.

Hunt, J.A., 1942, Geology of the Zeila Plain, British Somaliland: Geol. Mag., Vol. 79, pp. 197–201.

Kronberg, P., Schonfeld, M., Gunther, R., and Tsombos, P., 1974, ERTS-l Data on the Geology & Tectonics of the Afar/Ethiopia & Adjacent Regions, in Pilger, A., and Rosier, A. (Eds.): Afar Depression of Ethiopia: Proceedings of an International Symposium on the Afar Region and Related Rift Problems, Vol. 1, Inter-Union Commission on Geodynamics Sci. Rep. No. 14, pp. 19–26.

Larsson, I., Groundwater Investigations in a Granite Area of Sardinia, Italy, UNESCO, 1984, pp. 189–195.

MacFadyen, M.A., 1950, Sandy Gypsum Crystals From Berbera, British Somaliland: Geol. Mag., Vol. LXXXVII, No. 6, pp. 409–420.

MacFadyen, M.A., 1958, Igneous Rocks at Gorei, Shilemadu Range, British Somaliland: Geol. Mag., Vol. XCV, No. 3, pp. 172–173.

Merla et al., 1979, a Geological Map of Ethiopia and Somalia (1973) and Comment with Map of Major Landforms, 1:2,000,000, Consiglio Delle Richerche, Italy.

McCuen, R.H., 1982, A Guide to Hydrologic Analysis Using SCS Methods, Prentice Hall, Englewood Cliffs, N.J.

Morton, W.H., and Black, R., 1975, Crustal Attenuation in Afar, in Pilger, A., and Rosier, A. (Eds.): Afar Depression of Ethiopia: Proceedings of an International Symposium on the Afar Region and Related Rift Problems, Vol. 1, Inter-Union Commission on Geodynamics Sci. Rep. No. 14, pp. 55–65.

National Climatic Data Center, 1974, Summary of Synoptic Meterorological Observations, East African & Selected Island Coastal Marine Areas: Asheville, N.C.

Nelson, H.D. (Ed.), 1982, Somalia, A Country study: American University, Washington, D.C.

Pallister, J.W., 1958, Mineral Resources of Somaliland Protectorate: Overseas Geology and Mineral Resources, Vol. 7, No. 2, pp. 154–165.

Parkinson, J., 1932, A Preliminary Note on the Borama Schists, British Somaliland: Geol. Mag., Vol. 69, pp. 517–520.

Popov, A., Kidwai, A.L., and Karani, S.A., 1971, Ground Water in the Somali Democratic Republic, in: The Role of Hydrology & Hydrometeorology in the Economic Development of Africa, Vol. 2, Technical Papers Presented to the EDA/WMO Conference, Addis Ababa, 13–23 September 1971, WMO Rep. No. 301, World Meteorological Organization, Geneva, pp. 340–342.

Rodier, J.A. (Ed.), 1982, Estimation of the Annual Runoff of Small Basins of Sahelian'and Sudano-Sahelian Areas of Africa: International Hydrological Programme, Studies and Reports in Hydrology, Vol. 32, Gentbrugge, pp. 57–81.

Ruegg, J.C., Lepine, J.C., Tarantola, A., and Leveque, J.J., 1981, The Somalian Earthquakes of May 1980: Geophysical Research Letters, Vol. 8, No. 4, pp. 317–320.

Salad, M.K., 1977, Somalia, A Bibliographical Survey: Greenwood Press, Westport, CT, pp. 64–69, 232–315.

Schaefer, H.U., 1975, Investigations on Crustal Spreading in Southern and Central Afar (Ethiopia), in Pilger, A., and Rosier, A. (Eds.): Afar Depression of Ethiopia: Proceedings on an International Symposium on the Afar Region and Related Rift Problems, Vol. 1, Inter-Union Commission on Geodynamics Sci. Rep. No. 14, pp. 289–296.

Sogreah Consulting Engineers (Grenoble, France), 1981A, Northwest Region Agricultural Development Project Feasibility Study and Technical Assistance, Technical Report No. 1, Geomorphology: Somali Ministry of Agriculture, Mogadishu.

Sogreah Consulting Engineers, 1981B, Northwest Region Agricultural Development Project Feasibility Study and Technical Assistance, Technical Report No. 2, Climatology: Somali Ministry of Agriculture, Mogadishu.

Sogreah Consulting Engineers, 1981C, Northwest Region Agricultural Development Project Feasibility Study and Technical Assistance, Technical Report No. 3, Hydrology: Somali Ministry of Agriculture, Mogadishu.

Sogreah Consulting Engineers, 1981D, Northwest Region Agricultural Development Project Feasibility Study and Technical Assistance, Technical Report No. 5, Soil Survey: Somali Ministry of Agriculture, Mogadishu.

Sogreah Consulting Engineers, 1981E, Northwest Region Agricultural Development Project Feasibility Study and Technical Assistance, Technical Report No. 8, Soil & Water Conservation: Somali Ministry of Agriculture, Mogadishu.

Sogreah Consulting Engineers, 1981F, Northwest Region Agricultural Development Project Feasibility Study and Technical Assistance, Technical Report No. l6, Hydrogeology (Final Report): Somali Ministry of Agriculture.

Sogreah Consulting Engineers, 1982, Northwest Region Development Project Feasibility Study and Technical Assistance, A—Rainfed Agriculture, Vol. 1, Agricultural Development: Somali Ministry of Agriculture, Mogadishu.

Sogreah Consulting Engineers, 1982B, Northwest Region Development Project Feasibility Study and Technical Assistance, A—Rainfed Agriculture, Vol. 2, Soil & Water Conservation Works: Somali Ministry of Agriculture, Mogadishu.

Somaliland Oil Exploration Co., 1954, A Geological Reconnaissance of the Sedimentary Deposits of the Protectorate of British Somaliland: Crown Agents, London.

Styles, P., and Hall, S.A., 1980, A Comparison of the Seafloor Spreading Histories of the Western Gulf of Aden and the Central Red Sea, in Carelli, A. (Ed.): Geodynamic Evolution of the Afro-Arabian Rift System, Accad. Naz. dei Lincei, No. 47, Rome, pp. 585–606.

Tazieff, Varet, J., Barberi, F., and Giglia, G., 1972, Tectonic Significance of the Afar Depression: Nature, Vol. 235, No. 5334, pp. 144–147.

Vitale, C.S., 1967, Bibliography of the Climate of the Somali Republic: U.S. Department of Commerce, Washington, D.C.

Winch, K.L. (Ed.), 1976, International Maps and Atlases in Print (2nd ed.): Bowker, London.

MAP REFERENCES

Defense Mapping Agency, 1978, Series 2201, Djibouti, Sheet #21 (ed. 5), 1: 2,000,000: Washington, D.C., and Ministry of Defense, 1983, Tactical Pilotage Chart K6-D (ed. 4), Ethiopia, French Territory of the Afars and Issas, Somalia Democratic Republic, 1:500,000: Washington D.C. and HMSO, London.

Hoggaanka Kartografiyada Wasaaradda Gaashaandhigga J.D.S.

Defense Ministry Cartographic Directorate, S.D.R., 1976A, Jamhuuriyada Demuqraadiga Soornaaliyeed (Somali Democratic Republic), 1:200,000: Defense Ministry Cartographic Directorate, S.D.R.

Defense Ministry Cartographic Directorate, S.D.R., 1976B, Jamhuuriyada Demuqraadiga Soomaaliyeed (Somali Democratic Republic), 1:100,000: Defense Ministry Cartographic Directorate, S.D.R.

Directorate of Colonial Surveys, 1952–59, Maps of Somaliland Protectorate, Series DCS 39 (DOS 539), 1:125,000: Ordnance Survey, Southampton, U.K.

Directorate of Overseas Surveys, 1957–73, Geologic Maps of Somaliland Protectorate, Series DOS 1076, 1:125,000: Ordnance Survey, _outham_ton, U.K.

Merla et al., 1979, A Geologic Map of Ethiopia and Somalia, 1973, and Comment with a Map of Major Landforms, 1:2,000,000: Consiglio Nazionale Delle Ricerche, Italy, Pergamon Press, New York.

4 Sudan Case Studies and Model

INTRODUCTION

Over a three-year period of deep bedrock groundwater exploration in the Sudan in the late 1980s, a team of American scientists carried out two studies under contracts with USAID using innovative groundwater exploration technologies and modern geological theory that led to the discovery of regional, deep bedrock groundwater resources pervading the Sudan and its neighboring states, introducing the term "megawatersheds" into the lexicon of modern hydrology. By the 1990 conclusion of the second contract, 14 favorable groundwater development areas and 23 water-based economic development opportunities were identified in Sudan's Red Sea Province. Three "pilot" groundwater development projects were planned and ready for implementation. The three pilot project projects planned near the city of Port Sudan had near-term, incremental development potential of up to 10 million gallons per day of potable water. Overall, newly discovered megawatersheds in the Red Sea Province had the potential to supply urban areas, mining, and agricultural business with tens of million of gallons per day of useable water from local, cost-effective sources.

The unprecedented hydrogeological and geophysical data acquired during the Red Sea Province megawatersheds exploration program led to the first publication of scientific papers describing the Megawatersheds Model in 1989 [Bisson and El-Baz; Third World Academy of Sciences] and in 1990 [Bisson and El-Baz; Environmental Research Institute of Michigan (ERIM) 23rd Symposium of Remote Sensing of the Environment Proceedings]. The original megawatersheds model, as presented in Trieste in 1989, is included at the end of this chapter.

Modern Groundwater Exploration: Discovering New Water Resources in Consolidated Rocks Using Innovative Hydrogeologic Concepts, Exploration, Drilling, Aquifer Testing, and Management Methods, by Robert A. Bisson and Jay H. Lehr
ISBN 0-471-06460-2   Copyright © 2004 John Wiley & Sons, Inc.

Setting and Background

The Democratic Republic of Sudan is the largest country in Africa, arid and semi-arid in the Northern two-thirds and with swamps, tropical forests, and savannah in the South. The most influential natural feature of the Sudan is the Nile and its tributaries, where people, cultures, and trade have thrived since before recorded history. The Sudan borders eight neighboring countries, but its culture and political fortunes have been most closely linked to Egypt. In modern times, perhaps the most important underlying factor in Egyptian-Sudanese political relations is Egypt's utter dependency on continued access to a large fraction of the Nile's flow through the Sudan. In fact, Egypt possesses superior water rights to the Nile by agreements of 1929 and 1959 that severely constrain Northern Sudan's economic development potential in the face of the advancing desertification of the region, while neighboring Ethiopia was pursuing plans for massive dam construction projects on the Blue Nile and its tributaries that could considerably reduce the Blue Nile flows where it joins the White Nile in Khartoum and endanger Khartoum's agriculture-dependent economy.

NORTHERN SUDAN—1987

The USAID mission in Khartoum engaged the services of BCI-Geonetics, Inc. to perform a preliminary regional study of northern Sudan (Figure 14) to determine the feasibility of discovering large amounts of previously undeveloped, renewable groundwater. Over a five-month period, senior explorationists from BCI applied new concepts and technologies to re-examine Sudan's vast, diverse landscape, including thousands of square kilometers of desert, mountains, wetlands and coastal plains, and the greatest of Africa's rivers the River Nile, with its main tributaries the Blue and White Niles, draining vast areas of high rainfall in neighboring Ethiopia, Uganda, and the Central African Republic.

This initial aerial and ground reconnaissance of the Sudan's complex geomorphology by a BCI team was combined with interpretation of aerial and satellite imagery, topographical and geological maps, and literature review and interviews with Sudanese groundwater experts (see Acknowledgments). Using all of this information, the BCI team reinterpreted the Sudan's groundwater hydrology in context with regional tectonic fabric and rainfall patterns to produce a modern hydrogeological depiction of a groundwater resource of regional pro-

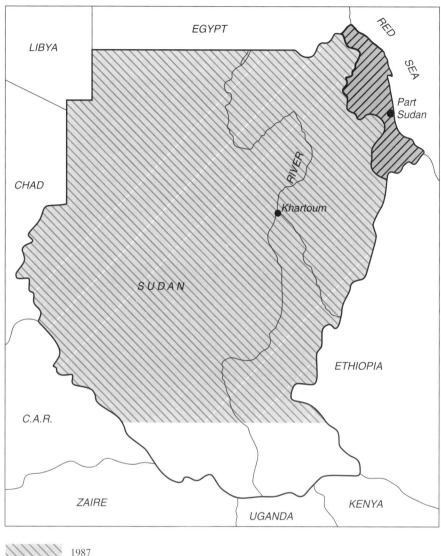

Figure 14. Locations of megawatershed exploration project areas.

portions flowing through complex open fracture networks underlying both humid and arid areas of the Sudan and its neighboring states.

The most important result of this 1987 feasibility study was discovering evidence of deep, fractured bedrock groundwater systems potentially connecting the Sudan to previously unknown groundwater recharge areas both inside and outside the country and the creation by BCI's team leader, Robert Bisson, of a conceptual model describing these regional deep groundwater flow systems. Bisson named the model "Megawatersheds" because the bedrock catchments appeared to transcend the local topographic constraints of surface watersheds and have the potential to carry deep groundwater far away from high rainfall areas to remote desert or offshore discharge. The model accounted for hydrological paradoxes and "anomalies" with far-flung, desert high-yield bedrock wells and springs and to provide new insights into the poorly understood groundwater hydraulics of vast areas of North and East Africa, from the great Sud wetland of the Sudan's Southern provinces to the Kufra Basin and Qattara Depression of Libya and Egypt.

The 1987 BCI study indicated new opportunities existed for economic development in arid areas such as the Sudan's Northern Kordufan and Red Sea Provinces. The USAID Khartoum Mission evaluated this new information and shared the results with Sudan's government and private sector.

RED SEA PROVINCE—1988–1990

In 1988, the Sudanese Company for Water and Economic Development (SUWED) applied for and was subsequently awarded a grant from the United States Agency for International Development (USAID) to establish the feasibility of developing water-dependent economic resources in the Red Sea Province of Sudan (Figure 15). The public-private sector initiative was based on BCI's 1987 study of fractured basement groundwater sources.

The ultimate objective of the feasibility program was to assess the physical and financial viability of project-specific water-related economic development in the Red Sea region and then act on that knowledge by drilling water production wells and building new businesses, bringing economic prosperity to the region.

In their agreement with USAID, SUWED principals were formally committed to investing in business development tied to mining, agriculture, and urban industry within the Red Sea Province once water development feasibility was proven.

The Feasibility & Pilot Project Design Program was accomplished by April 1990, identifying three pilot project areas for water well drilling and specific economic development projects, with several test

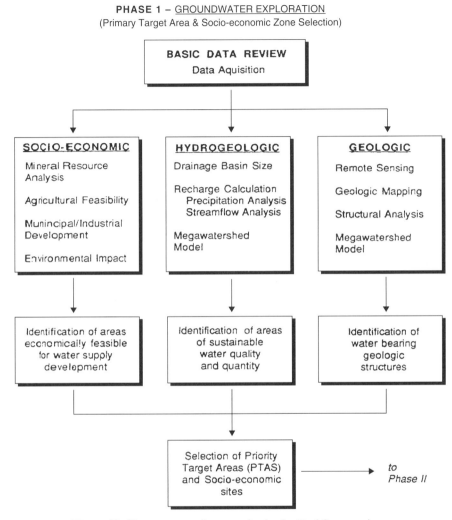

PHASE 1 – <u>GROUNDWATER EXPLORATION</u>
(Primary Target Area & Socio-economic Zone Selection)

BASIC DATA REVIEW
Data Aquisition

SOCIO-ECONOMIC

Mineral Resource
Analysis

Agricultural Feasibility

Munincipal/Industrial
Development

Environmental Impact

HYDROGEOLOGIC

Drainage Basin Size

Recharge Calculation
  Precipitation Analysis
  Streamflow Analysis

Megawatershed
Model

GEOLOGIC

Remote Sensing

Geologic Mapping

Structural Analysis

Megawatershed
Model

Identification of areas
economically feasible
for water supply
development

Identification of areas
of sustainable
water quality
and quantity

Identification of
water bearing
geologic
structures

Selection of Priority
Target Areas (PTAS)
and Socio-economic
sites

*to
Phase II*

Figure 15. New water and prosperity in the Red Sea province.

well sites actually surveyed, and with access and drilling permits obtained before BCI was asked by USAID to cease work on the project due to U.S. law prohibiting further funding in the Sudan because the new Sudan Government was not paying its debt service to the U.S. Government.

The project assessed groundwater development potential and economic development feasibility. The project report and test-drilling plan, completed in 1990, described 23 promising economic development sites and 14 new water development zones. Phase I groundwater investigations supported the "Megawatersheds" concept, presenting substantial correlations between megashear/fault zones and basin drainage within the Red Sea Province and new conclusions regarding coastal mountain catchments contributing active recharge to fractured bedrock groundwater systems pervading the region. Prior investigations of the potential water resources within the Red Sea Province had not recognized these correlations or their contributions to the potential discovery of major sustainable yields of usable water from bedrock sources.

The economic development potential was evaluated for several sites where water-constrained industries were proximal to newly discovered megawatersheds, including cement manufacturing; copper, zinc, and gold mining operations; agricultural development; and municipal water development for industrial and human consumption. With the successful development of these potential new water sources, economic development in Sudan's Red Sea Province would be considerably enhanced.

Hydrogeological and Geological Studies

SUWED received a grant from USAID to establish the feasibility of developing water-dependent economic resources in the Red Sea region of Sudan based on the discovery of new groundwater sources. Once feasibility was proven, SUWED principals agreed to invest in development of new groundwater supplies and new ventures associated with these water sources within the Red Sea Province, including mining, agriculture, industry, and municipal development.

Economic Potential of Sudan's Red Sea Province

Minerals   The mineral development potential of the Red Sea Region was quite favorable, but it hinged on the sustainable availability of reliable, inexpensive fresh or brackish water. The development of metallic minerals, including copper, zinc, and gold was promising within the western edge of the Red Sea Region. Potentially economic sources of industrial minerals such as cement and plaster have also been located within the Red Sea Region. Some of the factory equipment used in processing these materials were imported but not used due to the non-availability of affordable water.

Agriculture   The Red Sea Province contains a potentially profitable agricultural region within the Tokar Delta, with only 40% of the fertile lands within the delta used for agriculture. The many unknown and unpredictable aspects of both surface and groundwater in the Tokar region hindered the development of profitable agribusinesses.

Industrial and Municipal   There was a great demand for potable fresh water to supply both industry and residents of the City of Port Sudan and other coastal towns. Discovery of water near Port Sudan would not only aid the overexpanding population, but also encourage further expansion of existing businesses.

Overview of Work

The scope of work completed in Phase I is described on the accompanying flow chart (Figure 15). The interpretation and synthesis of hydrogeological, geological, and socioeconomic data resulted in the delineation of 14 Priority Target Areas (PTAs) for detailed Phase II hydrogeological studies and 23 favorable sites for Phase II assessment of economic development potential.

The 14 PTAs identified by BCI represented the areas most likely to contain sustainable, high quality and quantity water resources. The PTAs were divided into two categories: Primary Target Areas and Secondary Target Areas. The 23 socioeconomic favorable sites were also ranked using a matrix analysis. In addition, a preliminary hydrogeological analysis was performed on each of the 23 favorable socioeconomic sites.

During Phase II, BCI provided SUWED and USAID with a detailed synthesis of this information, including the five highest priority areas for groundwater development and a map of the Red Sea Project Area showing the location of Priority (groundwater) Target Areas and the socioeconomic Target Areas in context with topography, rivers, towns, railroads, and roads within the project area.

Site Selection Process—Red Sea Province Megawatershed Exploration Model

The groundwater exploration concept applied to the Red Sea study area was developed by BCI principals and coined the Megawatershed Exploration Model by BCI's founder, Robert Bisson. Salient characteristics of the Sudan Megawatershed Model are described later in this chapter.

The Red Sea Province contained several areas that fit this "Megawatersheds" model, especially the Tokar (PTA 5) and Port Sudan

(PTA 4) megawatersheds. These areas contain large basins dissected by numerous complex sheer zones. These sheer zones are areas that contain an abundance of fractured bedrock. The preliminary mega-watershed exploration model presented here is generic. The hydrogeological variations in the Red Sea Region that control potential water availability were analyzed in detail for each target area during the project and reported to the Sudanese Company for Water and Economic Development and USAID. Several well sites were surveyed with geophysical instruments, and wells were targeted for drilling. Drilling methods were recommended, and technical terms of reference for tender documents were prepared by the BCI Team with the assistance of U.S. drilling experts from the Sergent Drilling Company.

In 1990, USAID was compelled by U.S. law to terminate the planned drilling phase due to intergovernmental difficulties between the new Government of Sudan and the United States.

Conclusions

(1) The correlations among megashear zones, faults, large catchment areas, and subsurface drainage in the Red Sea Province are overwhelming. Prior investigations of the potential water reserves within the Red Sea Hills area have not emphasized these correlations or their contributions to the discovery of sustainable yields of usable water. The Megawatershed Exploration Model serves as the framework for a scientific analysis of these correlations.

(2) The successful demonstration of water and economic development feasibility during this pilot project should eventually trigger successful business ventures in the region. The results will lead to an substantially increased economic and social prosperity for communities in the Red Sea region of the Sudan.

MEGAWATERSHEDS MODEL—1989*

Introduction

The Megawatershed Exploration Model was originally developed on the basis of discoveries in Africa rift-related regional fracture

---

* The Red Sea Province Megawatershed Model was originally drafted by Robert A. Bisson, then chairman of BCI-Geonetics, and Dr. Farouk El-Baz of Boston University for presentation at the Third World Academy of Sciences Conference on Desertification in Trieste (1989).

permeabilities as a conceptual tool to enhance success in groundwater exploration, as an exploration model rather than as a quantitative hydrogeological model. However, after completing a substantial portion of the Red Sea Province exploration program, BCI's team leader Robert Bisson collaborated with senior advisor Dr. Farouk El-Baz in an attempt to apply the model to quantifying groundwater potential and then comparing the amount of potential resource with traditional watershed models. The results showed a marked increase in renewable groundwater availability in the megawatershed model calibration over traditional models and encouraged Bisson to lead his company forward in further research and development of the megawatershed paradigm to the present time.

The term megawatershed in this context describes the broadest possible three-dimensional catchment area and transmitter of water, originating from one or more recharge zones, and with surface and subsurface flow strongly controlled by regional structural features in the bedrock. Some of these structural features act as zones of secondary permeability, through which large amounts of water may flow (Figure 16). In this case, *mega* refers more to order-of-magnitude effects on groundwater flow in these three-dimensional fracture controlled systems than to areal basin size.

The purpose of the model is to better define these order-of-magnitude effects to assist in the discovery of large amounts of high-quality groundwater, here termed "high-grade water" located within zones of secondary permeability. The "*high-grade water*" is attained when the measured yield from a well is an order of magnitude greater or more than the average yield of other wells in the region.

Background   Large regional aquifer systems occur in many parts of the world. In many cases, the subsurface movement of water over long distances is a necessary ingredient of conceptual models of the dynamics of such aquifers. Most of the known regional systems, such as The Great Artesian Basin of Australia, the chalk aquifers of France and England, the Nubian Sandstone systems of North Africa, and the Ogallala Formation of the High Plans Region of the United States, are generally believed to occur entirely within sedimentary rock formations.

Also generally accepted is the view that aquifers in these sedimentary rock formations derive their characteristics from the imposition of climatic and hydrometeorological conditions upon the regional geology. With the exception of certain limestone terrains, primary properties of rock permeability and the geometry of associated sedi-

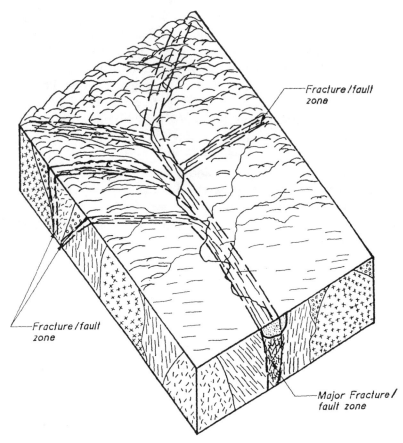

Figure 16. Schematic illustration of the control drainage by major fault/fracture zone.

mentary formations were in the past seen to represent the only major parameters affecting the occurrence of groundwater in sedimentary basins.

The premise of the megawatershed model is that, in addition to large watersheds underlain by sedimentary rocks, where primary porosity is assumed to control groundwater movement, there are also regional basins consisting of interconnected fracture systems (Figure 16). These can occur within watersheds underlain by both crystalline and sedimentary rocks.

Formal studies of groundwater behavior related to bedrock fractures have been carried out for more than a century. As early as 1835, William Hopkins, in *Researches in Physical Geology*, listed his observation of rectangular arrangements in topographic features, drainage patterns, faults, mineralized veins, joints, and alignment of springs. Dana [1847] contended that controlled reestablishment of the regional fracture system takes place along preexisting patterned stress field lines, later confirmed by Cloos [1948] and Meinesz [1947]. Hobbs [1911] stated that the development of land forms, in some areas, is largely the result of systematic bedrock fracture control and that these fracture systems propagate and manifest themselves on the Earth's surface in the form of drainage alignments, linear topographic trends, soil-tonal anomalies, and vegetation alignments, later confirmed by Hilgenberg [1949], Hill [1956], Henderson [1960], and Badgely [1965].

Historically, the propagation force has been attributed to Earth tides, isostatic rebound, active crustal stress patterns, ocean floor spreading, and differential compaction over an irregular basement. Casagrande [Mollard, 1957] indicated that a greater hydraulic gradient exists in the overburden above a fractured zone of bedrock. Downward movement of groundwater will cause a leaching of fine-grained particles and consequent subsidence. The fractured zone, under certain conditions, may act as a vertical conduit for discharging groundwater, accounting for higher soil moisture and corresponding vegetation alignments. Angelillo [1959] in his regional analysis of the Mojave River basin groundwater regime graphically described a stress-field-induced and fractured-rock-controlled basin recharge and discharge system.

The advent of modern Earth orbital platforms with remote-sensing instrument payloads, in the 1970s, combined with a concurrent wide acceptance of applied plate-tectonic theory, encouraged a new genera-tion of explorationists to reexamine existing concepts in a new light. In recent studies of aquifer systems, for example, Buckley and Zeil's [1984] study on eastern Botswana, the role of secondary permeability is seen to be an important part of the aquifer hydrodynamics.

Increasing attention is being given to the superimposed secondary effects of structural and tectonic influences on the primary hydraulic properties of the rock formations. For example, the hydrogeology of the Triassic sandstones of central England [Price, 1985], the chalk of eastern England [Nunn et al., 1983], and the Tertiary limestones of Florida [Cederstrom et al., 1983] are now known to be strongly influ-enced by faults and their associated fracture systems. In another example, Zacharias and Brutsaert [1988] show that although the recharge characterization of a watershed as a single unit may be a useful approximation, it can also have its limitations.

The total groundwater discharge from a watershed is the sum of flow contributions from various sections along the aquifer such as infiltra-tion at the head waters, transmission, and storage along the channel-ways at the discharge areas, each of which have unequal behavioral properties (i.e., the steeper parts of the drainage basin behave differently from those that are more gentle). Results of their studies show that secondary permeability greatly increases the total watershed base flow and the recharge calculations.

Model Components   In order to explain the flow system of a particular aquifer with respect to an entire watershed, it is common to develop a simple conceptual model, which describes the physical conditions of each of the components of the hydrogeological cycle. This includes a description of the source of the water (usually precipitation), how the water moves to and within the watershed (infiltration and transmission), how the water is stored within the aquifer, and the nature of discharge from the system.

In the case of the megawatershed model, in addition to an evaluation of primary aquifer characteristics, results from descriptive site examinations, laboratory studies, and computer modeling of fluid flow in fractured rock are also used to document the behavior of groundwater in all parts of the watershed.

Water Source   Although many exceptions exist, such as the potential Okavango Delta recharge of a Kalahari Megawatershed [Bisson, 1985], in most cases, groundwater has its direct source in precipitation. In many arid regions, the average lowland or basin rainfall is usually low and the evaporation rate high. For example, in Libya, rainfall varies from less than 25 to 400 mm per year, with potential evaporation rates at 1700 to 6100 mrr! per year; yet flow from a well field adjacent to springs in the Libyan desert has been measured at 250,000 cubic meters per day (45,000 gpm). In another case, La Moreaux et al. [1985] in their study of the Kharga Oases in the western desert of Egypt, found a mean annual rainfall of 1 mm, with potential evaporation rates similar to those cited above. Yet individual wells at the oases flowed at more than 15,000 cubic meters per day (2700 gpm). Elsewhere, similar regimes of low rainfall, high potential evaporation, and anomalously productive springs have been noted as in the Sinai Desert [Issar, 1985].

Studies of some of these areas have stated that local recharge cannot account for the measured discharge [Abufila, 1984; Ahmad, 1983]. One possible answer is that recharge is coming from further away, in mountainous terrain. The "orographic effects" of mountainous terrains on rainfall can induce precipitation amounts, which are considerably greater than in lowland regions (Figure 17). Given the distinct possibility that this meteoric water might enter subsurface fracture conduits

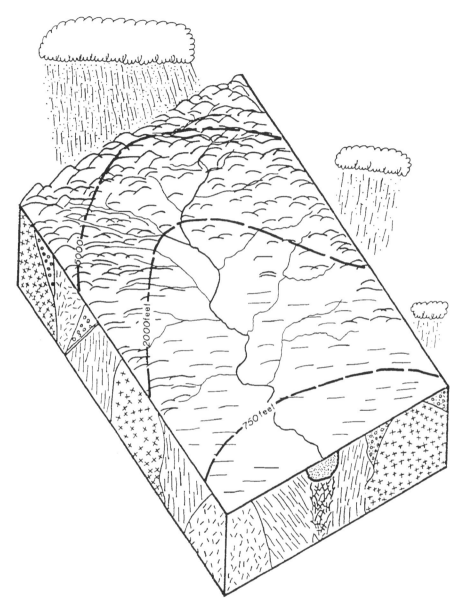

Figure 17. Geographic effects on precipitation caused by topographic variations in the terrain.

connecting the mountains to the distant desert springs, mountains may be a reasonable recharge source to consider. For example, the potential groundwater recharge rate from the Chad and Egyptian mountains to the Kufra and Sarir well fields (600 miles away) has been estimated at more than 70 million cubic meters per day [Alam, 1989]. That is considerably more than the two million cubic meters per day drawn from the well fields.

In some areas, where elevations and climate dictate, precipitation may also occur as snow. Angelillo [1959] suggests that the snow pack may increase infiltration by decreasing rapid runoff (except during rapid spring melts). The times of slower melting would allow for constant subsaturation of the porous media below the snow and thus increase the infiltration rates. This would also increase the amount of water available for recharge.

It is also well established that over the past 200,000 years, the climate in many arid regions was different, with local rainfall rising to 1000 mm instead of the current 100 mm or less. Alternating wet and dry climates were the rule during the Holocene [El-Baz, 1989]. Geoarcheological evidence indicates that the North African Sahara enjoyed wet climates 4000 to 11,000 years ago, 20,000 to 35,000 years ago, about 60,000 years ago, about 140,000 years ago, and about 210,000 years ago. Rainfall during these periods would have been enough to support much vegetation and animal as well as human populations. Much of that water, which created river courses and vast lakes [El-Baz, 1988], would have been stored in groundwater aquifers.

Therefore, some of the water stored in the present day aquifers is likely to be paleowater. Issar [1985] found that the water from large springs in the Sinai Desert was 20,000 to 30,000 years old. La Moreaux et al. [1985] proposed the same for water found at the Kharga Oases in Egypt. The apparent age of such waters may also be due to the great distances that some of the water at these oases might have traveled (tens to hundreds of kilometers), along subsurface conduits, between source and spring [Abufila, 1984; Ahmad, 1983]. The very slow natural velocity of water through an aquifer results in a major age difference along the length of the system. In effect, the age of the water coming out of desert springs is much older than the groundwater near the source of recharge.

We caution against establishing the presumption that recharge is not sufficient, or that the "paleowater" discovered in major desert basins will be "mined out" and not replenished. Before such a conclusion is made, the calculation of potential recharge must include more measurements of the possible increase in recharge due to (1) spatial and temporal variations in precipitation versus the very limited baseline precipitation data of remote arid environments; (2) the effects of increased infiltration associated with zones of secondary permeability in mountainous areas; and (3) the effects of regional, increased transmissivity from enhanced secondary permeability imparted by fractures and faults. In addition, the induced infiltration effects on long-term, large-scale pumping of wells must be considered as a probable outcome in the development of these resources.

Infiltration   An important part of recharge estimation is related to the size of the catchment area, or "watershed," and the amount of water that is estimated to infiltrate into groundwater conduits. It is obvious that large catchment areas (>20,000 km^2) have the potential to yield more water for recharge (Figure 18). However, smaller basins (<1000 km^2), in some cases, also have the potential to yield high-grade water.

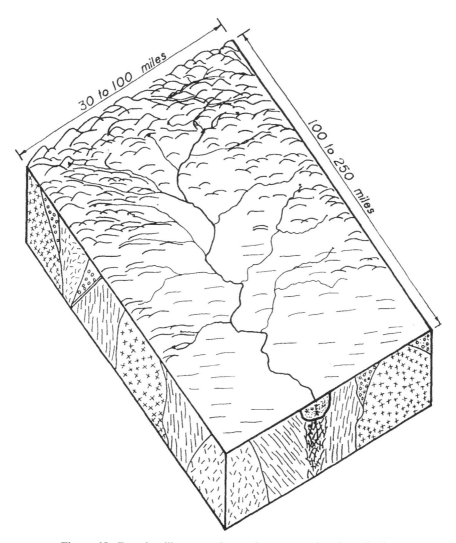

Figure 18. Drawing illustrates the catchment area in a large basin.

The potential for high-grade water yield is related to the permeability of materials directly underlying principal rainfall areas. This surface permeability strongly influences infiltration rates to unconsolidated and consolidated formations, with varying storativities and transmissivities, related both to primary and to secondary attributes of permeability.

Precipitation magnitude, an equally important factor, is spatially highly variable, often due to local terrain (elevation) effects. When natural conditions result in highly permeable materials (especially, fractured rock surfaces) being subjected to heavy precipitation events, as in many arid mountainous areas, the condition exists for substantial actual infiltration amounts in spite of high "potential evaporation" or other theoretical impediments to groundwater recharge. The combination of these "remote-recharge" factors may account for the major fraction of available new water for arid areas megawatersheds.

Of course, infiltration rates will vary with those physical and chemical qualities that control runoff and evaporation, including surface texture, porosity, degree of soil saturation, amount/type of vegetation, and slope. For example, if the surface texture is smooth with a low permeability, then runoff will be high and the amount of water that infiltrates will be small, even if the watershed is large. The opposite is also true.

In basins that contain fracture/fault zones (Figure 19), the infiltration rates are likely to be extremely variable. Rocks that are cut by these fracture/fault zones erode faster, leaving a highly irregular surface with large, polygonal, tubular, and disk-shaped voids. The largest frequency of these surface voids is generally adjacent to and within brittle fault zones. At some distance from these zones, the density decreases, the surface becomes much smoother, and the porosity decreases rapidly; so does the infiltration rate.

Often there is some interconnected permeability in the fractured zones with a much lower permeability in the surrounding rocks. It is possible that a low infiltration rate in the surrounding unfractured crystalline rock (<10 mm per day) exists along with a very high infiltration rate in the fractured rock (>440 mm per day), as similarly estimated by Chow [1964]. Fracture/fault zones often constitute local or regional topographic lows. This may result in a channeling effect, where water flows from smooth, impervious terrain into porous, fractured media.

These fracture zones of potentially high infiltration may occur in both nonporous (crystalline rocks) and porous media. The enhanced secondary porosity, rough surface texture, and increased transmissivity of fracture zones may cause increases in infiltration rates over the sur-

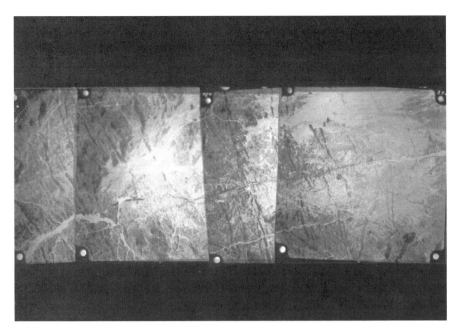

Figure 19. A strip of aerial photographs of fault-controlled drainage in the Red Sea province in Eastern Sudan. (See color insert.)

rounding host rock and allow more water to infiltrate into ground-water conduits. Assuming that the infiltration rates of the fracture zones are higher than those of the surrounding rock, the amount of water that is potentially available for recharge increases proportionally to the surface area of the exposed fracture zone. This effect may result in a multifold increase in the infiltration rate, over what would normally be calculated, without the effects of secondary permeability.

Transmission   In the megawatershed model, the subterranean faulted and fractured zones possess high infiltration rates and form three-dimensional polygonal drainage patterns with fractal properties and often have surface expression [Barker, 1988; Barbara and Rosso, 1989]. The fractal character means that small-scale fractured rock and drainage patterns are reflections of larger scale patterns that are similar in shape and orientation. Many of the smaller fractures observed in outcrop display shapes and orientations similar to those observed on a regional scale.

The smaller features that help to increase infiltration may be linked to the larger structural conduits, thus allowing a portion of rainfall that infiltrates at the headwaters to be transmitted along subsurface conduits. Eventually some portion of this subsurface water will exit (discharge) at a spring or well down gradient. The channeling of groundwater into subsurface conduits, associated with linear fault/ fracture zones, may often result in economic amounts of extractable groundwater.

Numerous studies have shown a positive relationship between fracture traces and lineaments and zones of increased well yields. Lueder and Simons [1962] presented some evidence that wells with relatively high yields in Alabama are found close to lineaments, whereas wells with medium to low yields are more randomly distributed. Sonderegger [1970] reported that in northern Alabama, wells located on fracture traces have yields four to five times that of wells located off of fracture traces. Siddiqui and Parizek [1971] showed that of 80 wells sampled in central Pennsylvania, those located on fracture traces, assuming a zone of fracturing 10 m across, had a median productivity 5.5 times greater than off-fracture trace wells, and 9.4 times greater yields than wells located at random. Cederstrom [1972], in relating drainage development to the regional fracture system in the Piedmont region, found that many valleys and valley intersections are fracture controlled [confirmed by Parizek, 1976]. He indicated that five wells located on major fracture valley intersections yield 1 mgd each.

Faust et al. [1984], in their analysis of the Baca geothermal system in New Mexico, calculated that water moved along fault systems and permeable horizons a distance of 20 miles and discharged along a fault at the Jemez springs. The calculated discharge was approximately 1.2 million gallons per day. Emery and Cook [1984] analyzed the results of test pumping a BCI Geonetics-sited production well in Connecticut, that intersected a fault zone in crystalline rock where the initial test yield was over 700 gallons per minute. Huntoon and Lundy [1979] found extremely high transmissivities associated with faults in the Casper aquifer in Wyoming. Discharges up to 1080 gallons per minute were measured along the faults. La Moreaux et al. [1985] measured a flow of 2750 gallons per minute along faults through the Kharga Oases in Egypt. Kohut et al. [1984] test pumped a well in fractured granitic rock at 250 gallons per minute and calculated a transmissivity of $6.3 \times 10^{-4} m^2/s$.) Smith [1980] found high flow and transmissivities associated with fault zones in Lincolnshire aquifer in England. Flow was measured at 3700 gallons per minute with transmissivities up to $6.9 m^2/sec$. In Italy, water flow from a fracture cutting a tunnel was measured at 9500 gpm (with 882 psi) and a total flow into the tunnel from several fractures was measured at 32,000 gpm [*World Water*, June

1989]. From these studies, it is clear that fracture/fault zones can transmit large quantities of water and in many cases contribute to the discovery of high-grade water.

Tsang and Tsang [1987] studied a 14,000 square meter area of crystalline rock in Sweden and found that 50% of the total groundwater flow was accounted for by 3% of the area due to the preferential flow of water along open fracture zones. In addition, there is preferential flow within the individual fracture system. Moreno et al. [1988] studied the flow of groundwater within a single vertical fracture system and found that only 10% to 20% of the fracture actually participated in the groundwater flow. They pointed out that specific conduits were developed along interconnected disk-shaped voids between areas of fill and wall contact. It may be likely that 50% of the groundwater flows through fracture conduits (three-dimensional braided network, with fractal properties) that occupy less than 1% of the volume of the watershed. This is a strongly anisotropic feature in the subsurface flow, and it has important ramifications with respect to groundwater exploration.

The anisotropy that is developed because of these preferential conduits strongly influences the behavior of groundwater in fractured terrain. As a result, zones containing high-grade water can occur. But the target aquifer represents perhaps 1% of the total volume of the host rock region and thus presents a challenge to the exploration hydrologist.

The superposition of increased infiltration associated with zones of secondary porosity, open fracture conduits (increased transmissivity), increased drainage density associated with fault/fracture zones, and an increased gradient can result in an increased groundwater flow and, thus, increased recharge. Smith [1980] proposes the concept of "rapid recharge" when such open fracture systems are a dominant part of the groundwater flow. His study in Lincolnshire, England, revealed fracture zones of rapid flow with extremely fast response times. However, it is possible that a fraction of the groundwater recharge is conveyed to the system almost "instantaneously" in order to account for the observed events.

If the above observations of fast response times and anomalously high yields in fracture zones (high-grade water) is reasonable, then we may assume some degree of fracture interconnectedness, along a reasonable gradient, from the zone of groundwater infiltration to the zone of groundwater discharge. The degree of the fracture interconnectedness and thus the flow through the subsurface system will vary due to many factors. The hydraulic conductivity of fracture systems is dependent on the density of the fractures, the aperture of the fracture, the degree of roughness, the tortuosity (ratio of traveled path over linear path), the interconnectedness of the fractures, and the hydraulic head. The number of rigorous studies on the fluid flow through highly fractured terrains over the length of a watershed is limited, but the following general observations can be made:

Fracture Characteristic	Effect on Flow
Fracture width-aperture	The permeability of smooth walled fractures is directly proportional to the cube of the fracture width (Sterns and Friedman, 1972)
Fracture density	Permeability increases as the fracture density increases (Sterns and Friedman, 1972)
Fracture connectedness	The larger the connectivity of the fracture network and intersection of fractures, the greater the permeability

The density of fractures needed for flow to be established is called the percolation threshold [deMarsily, 1985]. Throughout the length of a subsurface fracture conduit, many finite or discontinuous clusters of fracture zones would have to become connected at a level exceeding the percolation threshold. This would form a continuous fracture zone over the length of the watershed, with a relatively high transmissivity (Figure 20). In most fractured regions, the fracture density increases closer to a major structural weakness. As the fracture density increases or the ratio of joint length to joint spacing increases, so does the probability of fracture connectedness [Gale, 1982] and thus the probability of sufficient transmissivity.

Surface Roughness	The effects of roughness vary depending on the aperture. When the flow is nearly laminar, as within wide fractures, the effect is minimal [Moreno et al., 1988; Sterns and Friedman, 1972]
Tortuosity	The effects of tortuosity vary depending on the aperture. When the flow is nearly laminar, as within wide fractures, the effect is minimal [Tsang and Witherspoon, 1985].
Pressure Gradient/ Hydraulic Head	The more parallel the extensional fractures are to the gradient, the larger the permeability [Endo and Witherspoon, 1985; Sterns and Friedman, 1972; Gale, 1982].

The pressure is expressed as potential energy (e.g., pounds per square inch). The hydraulic head is defined at a point and is the measure of

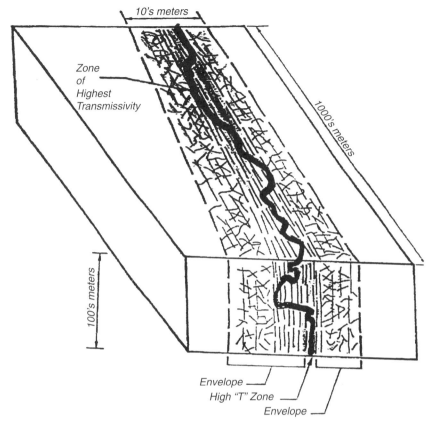

Figure 20. Detailed illustration of a fracture zone aquifer consisting of a highly transmissive zone and an envelope with high storativity.

groundwater level using a piezometer, compared with a reference pressure and elevation. The hydraulic head (or pressure) is composed of a gravitational head due to change in elevation and a pore pressure due to the confining pressure imparted on the pore fluid.

In near-surface fracture systems with wide apertures that are parallel to the gravitational gradient, it is likely that the pressure within that system will be dominated by gravity. This pressure will remain dominant until the confining pressure (of the rocks) exceeds the gravitational pressure (of the water).

In fractured, porous-medium aquifers at depth, the confining pressure (rock) may exceed the gravitational pressure (water). In these cases, the confining pressure of the surrounding rocks increases the hydraulic pressure within the fracture/fault systems that are transmitting large volumes of water through the basin as the pore fluid under pressure moves from the host to the fracture zone. The combination of increased confining pressure, increased gradient, and increased transmissivity could result in unusually high-yielding "artesian" wells that could endure for centuries.

In summary, the fractured rock conduit that is likely to transmit large amounts of water will:

- Have a high density of fractures
- Have fractures with wide apertures (extensional)
- Be parallel to gravitational gradient
- Receive adequate recharge
- Be overpressured

When these conditions coexist, the subsurface fracture zones will serve as conduits for the transmission of high-grade water along preferential zones that comprise a small percentage (one to three) of the total volume of material in the watershed. It is possible that the transmissivity along these conduits will be considerably higher (two to three times) than that normally calculated for permeable zones that do not have the overprint of a well-developed secondary porosity.

Storage   The total yield (specific yield) from an aquifer is proportional to the amount of water held in storage, the amount of water transmitted (pumped) out of storage, and the rate of water replenishment (recharge). Discussion of the characteristics of a fracture zones with high transmissivities also applies to the analysis of specific yield. A unique characteristic of fracture zones is that they may contain both a zone of high transmissivity with low storativity in the fracture conduit, and a zone of lower transmissivity with higher storativity enveloping the fracture conduit (Figure 20).

The previous discussion indicates that open fracture zones parallel to the hydraulic gradient are capable of transmitting large quantities of water and that they may also be recharged rapidly. This occurs within the fracture conduit. The fracture envelope around this conduit is less transmissive but serves as storage that replenishes the open conduit and is also replenished by this conduit (Figure 20). These characteristics of a continuous bedrock aquifer, composed of the fractured rock adjacent to a highly transmissive zone, either discreet or in combination with alluvium or some other porous medium (Figure 21),

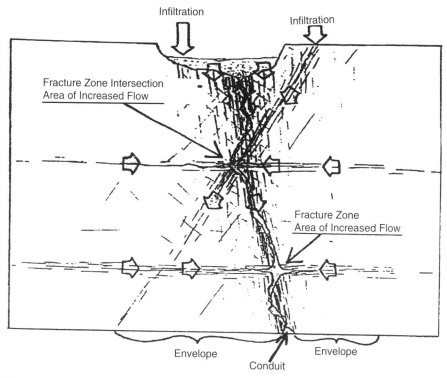

Figure 21. Cross-sectional view of a fracture zone aquifer illustrates the relation of flow to fracture intensity. Arrows show direction of groundwater flow. Dominant flow is perpendicular to the page.

define an important part of a megawatershed groundwater exploration target.

In addition, in the vicinity of the target aquifer, the fracture should have a moderate nondispersion factor. This means that the width of the aquifer and its associated conduits and micropathways is not so large as to disperse the amount of available recharge over a broad area composed of discontinuous aquifers. A large dispersion factor would disperse the available resource and decrease the potential yield of any single target aquifer. Conversely, a small in most cases defining the groundwater resource and its potential for long-term sustainable yield will require more sophisticated modeling than that applied to porous media. Some form of multiple porosity model may need to be applied due to the following:

(1) The conduit of the fracture zone will have porosity and transmissivity values that differ from the surrounding materials.

(2) The fracture zone envelope will have values significantly different from the host rock or the conduit.

(3) Both host rock and fracture zone porosity (conduit and envelope) will be significantly different from that of overlying sediments.

In such regions, the application of a double porosity model might be the best way to explain the flow characteristics. The open conduit would be characterized by a rapid flow and a rapid recharge, whereas the surrounding fracture envelope (and the alluvium) would be characterized by much lower transmissivity values but greater storativity.

For a fractured porous medium, Gale [1982] favors a "coupled discrete-fractured porous medium model" where faults and shear zones are defined as individual features. The permeability of the host rock is a feature and the joint characteristics another feature. For modeling a groundwater storage unit as illustrated in Figure 22, some modification of Gale's approach may be warranted. We propose using a discrete, individual characterization approach, for the conduit and the fracture zone, and an averaging or continuum approach for the host rock and the sediment. In addition, the model will define the interactions between media of differing porosity.

The model should be used to characterize the partial groundwater contributions to the total potential sustainable yield from each of the following:

(1) The sediment aquifer

(2) The conduit within bedrock

(3) The fracture envelope within bedrock

(4) The interplay among these three features with regard to the pumping rate and the recharge

Further research will shed light on the characteristics of various portions of these complex aquifer characteristics and how they may change as the hydrogeological parameters vary 1 and how these changes affect the potential sustainable yield of the resource.

From experience, we know that these fracture-controlled aquifers can exhibit large variations in quantities and qualities of water. The primary goal is to produce the greatest yield of the cleanest water with the lowest economic cost and environmental impact. The megawater-

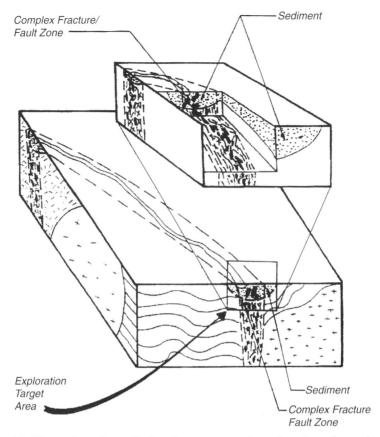

Figure 22. Illustration of an alluvium/fracture trap shows location of a major fracture/fault zone with alluvium acting as a water trap. Note the complex nature of the fault/fracture zone.

sheds model (which attempts to address the complexities of bedrock aquifer systems) provides groundwater explorationists with a qualitative perspective. Ongoing development and revisions of the model will better incorporate further on-site evaluations of watersheds and possibly include some computer modeling as data are compiled over time.

Groundwater Discharge   The concept of high-grade water associated with fracture zones is further supported by the occurrence of large fresh water springs at the distal ends of basins. These springs may occur on the coast, as in Libya, or they may occur in a rift valley as in the Sinai. As stated above, the flow rate in either case is anomalously high. The springs may also occur offshore. Large fresh water springs are known off the Mediterranean and Red Sea coasts of Africa. In many other coastal regions of the world, from the American coasts of Florida and California to the Arabian Gulf, large amounts of fresh and brackish water have been encountered. This suggests that there is a connection between basin edge occurrences of high-grade water and highly transmissive fracture zones that are connected to drainage basins. This is supported by the fact that high-grade water cannot be accounted for using standard local recharge and primary porosity models [Issar, 1985]. The high-grade water might be induced along zones of higher permeability, e.g., fault and fracture zones, and these zones might be connected to broader areas of recharge either through fractured rock alone or in combination with a rock unit possessing primary porosity.

Hypothetical Model Comparison Egypt/Sudan Red Sea Region

General Assumptions   The megawatershed model dictates that certain hydrogeological assumptions apply due to the effects of secondary porosity and permeability. The following comparison of three different models for a proposed well field, with natural environmental characteristics found in the narrow, coastal plain of the arid Red Sea Provinces of Sudan and Egypt, will help to illustrate these assumptions and the resultant order of magnitude difference of aquifer characteristics. The watershed example for these model comparisons is a 300-km^2 basin with 150-mm average rainfall (heaviest in the mountains), very little vegetation, soil or weathered rock cover, over an essentially exposed crystalline rock terrain. Alluvium is confined to "pockets" in the mountains, plus certain reaches of wadis and in alluvial fans between mountains and seacoast.

The three models are based on different hydrogeological and geologic assumptions. The general model assumptions are as follows:

Model 1:   Economically Accessible groundwater is stored only in alluvium in the coastal plain and fractured rock has a negligible effect on groundwater storage or flow. Recharge is limited to overland flows that extend beyond the mountains and infiltrate into coastal alluvium. Direct mountain recharge is not significant.

Model 2:   Economically Accessible groundwater is stored in alluvium and, locally (within 100 m of the well), in fracture zones. These local fractures contribute minimally to storage and, at best, may simply enhance local groundwater flow efficiency into bedrock well.

Model 3:   Economically Accessible groundwater storage and flows are strongly influenced by a regionally extensive network of subsurface fractures that extend essentially throughout the basin (Figure 23) and are largely in hydraulic contact with overlying alluvial deposits in the mountains, while possessing confined and semiconfined aquifer characteristics in the coastal plains. Recharge from mountain rainfall is assumed to contribute directly to this "megawatershed" (Figure 24).

Each of these models attempts to describe the nature of the aquifer connected to strategically located well fields. The values calculated for Models 1 and 3 are considered minimum and maximum values respectively, given the following:

Figure 23. Perspective illustration of the megawatershed model applied in the Red Sea province of Sudan. Arrows indicate overland and underground flow paths. (See color insert.)

Figure 24. Perspective illustration of the megawatershed model applied in the Red Sea province of Sudan. Arrows illustrate mountain recharge to fracture systems and alluvial storage. (See color insert.)

Catchment area	300 km^2
Average annual precipitation	150 mm
Total annual rainfall	45 million m^3

Wellfield Recharge Estimations   In this arid environment, a large part of the hypothetical 45 million cubic meters of annual rainfall evaporates or becomes part of the runoff associated with flash floods. Part of the precipitation infiltrates into the subsurface and is transmitted to the aquifer and to economically accessible well fields near the coast. Models 1 and 2 assume that the alluvium is the only place where significant infiltration, transmission, and storage can occur and that only local, coastal, rainfall, and overland flooding contribute to recharge. Because of the limited areal extent and thickness of the alluvium and the assumed minor role of local fracture networks in Models 1 and 2, the net percent of rainfall contributed to infiltration is low (1% to 3%).

Further, once the water becomes part of the subsurface, it must be transmitted to the aquifer at the target well fields, if it is to be considered part of the potential yield calculations. The alluvium is considered in all three models to be generally of uneven thickness and surficial groundwater is, in some locations, forced nearer to the surface causing large amounts to evaporate. In addition, some groundwater stays trapped within the alluvium and is not transmitted to the well sites.

Given the constraints of Model 1, estimates of 5% to 10% transmission of infiltrated water to the well sites are considered reasonable by the authors.

In Model 2, a slight increase of 0.5% is added due to the influence of local fractures.

In Model 3, the effects of secondary porosity and permeability over much of the catchment area result in considerable increases in the estimates of infiltration, transmission, and storage of groundwater. The numbers below illustrate this:

In Model 2, the alluvial storage remains as in Model 1, but it is supplemented by a local fracture system, consisting of two intersecting fracture zones, each 100 m long, 10 m wide, 100 m deep, and one horizontal fracture (200 m × 100 m × 1 m), with an average porosity of 0.05.

In Model 3, the alluvium is also considered as potential storage as in Models 1 and 2, but this is supplemented by a fracture zone network consisting of three vertical fracture zones extending nearly the length and width of the overall catchment area and one extensive horizontal fracture zone at depth. The width of each vertical zone is 20 m, depth 200 m, and combined length of 30 km. For purposes of simplicity, the considerable impact of the active mountain alluvial recharge and storage is ignored in this Model. The dimensions of the horizontal zone are 20 km × 5 km × 1 m. All fracture zones have an average porosity of 0.05.

Potential Storage Comparison [A = Alluvial F = Fracture zones]

	Volume (m^3)	Porosity	M^3	Total M^3
Model 1	A- 6.0×10^6	0.20	1.2×10^6	1.2×10^6
Model 2	A- 6.0×10^6	0.20	1.2×10^6	1.21×10^6
	F- 0.22×10^6	0.05		1.1×10^4
*Model 3	A- 6.0×10^6	0.20	1.2×10^6	9.2×10^6
	F- 160.0×10^6	0.05	8.0×10^6	

* Model 3 does not include contribution from mountain alluvium.

The above calculations predict the maximum amount of water that could be in storage under ideal conditions for each model. It is likely that both the alluvium and the fracture networks are not homogenous and are not hydraulically connected in a uniform manner. This means that a part, perhaps 50% of the groundwater in storage, is not available for easy extraction at the given well site. This results in potential yield calculations as follows:

Hydrogeological Parameters

	Infiltration	Transmission	% Rainfall	Cubic M/Year
Model 1	1–3%	5–10%	0.05–0.3%	2.25×10^4 to 1.35×10^5
Model 2	1–3%	5.5–10.5%	0.055–0.31%	2.4×10^4 to 1.4×10^5
Model 3	3–9%	10–30%	0.6–3.0%	2.7×10^5 to 1.35×10^6

The result of these calculations, expressed as average annual active recharge, are summarized below:

Potential Recharge Comparison

Model 1	370 m³/day	68 gpm
Model 2	384 m³/day	70 gpm
Model 3	3700 m³/day	678 gpm

From these hypothetical model comparisons, it is clear that Model 3 provides an order-of-magnitude larger amount of recharge than Models 1 or 2.

Storage Calculations   In all of these models, calculations of storage assume a continuous aquifer over which average values of effective porosity can be extrapolated.

The alluvial storage in Model 1 is limited to an already confined surface drainage, extending from the base of the mountains, 20 km downstream to the proposed well fields sites. The alluvium is considered to average 10 m in thickness and to average 30 m in width. The porosity is 0.2.

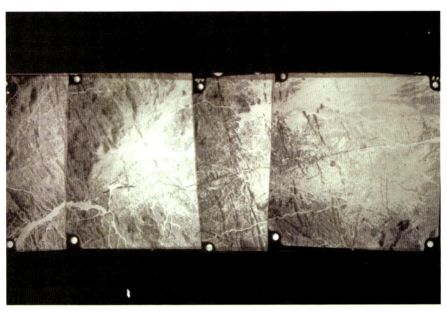

Figure 19. A strip of aerial photographs of fault-controlled drainage in the Red Sea province in Eastern Sudan.

Figure 23. Perspective illustration of the megawatershed model applied in the Red Sea province of Sudan. Arrows indicate overland and underground flow paths.

Figure 24. Perspective illustration of the megawatershed model applied in the Red Sea province of Sudan. Arrows illustrate mountain recharge to fracture systems and alluvial storage.

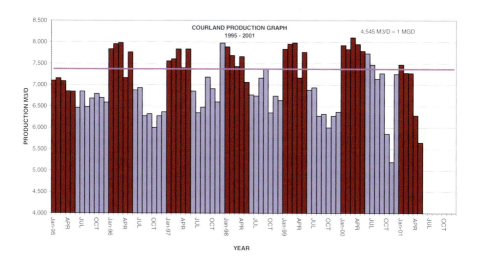

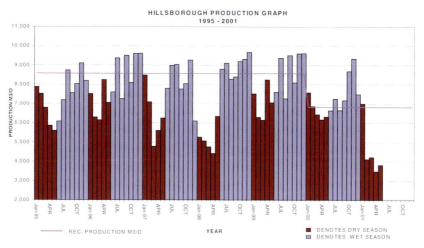

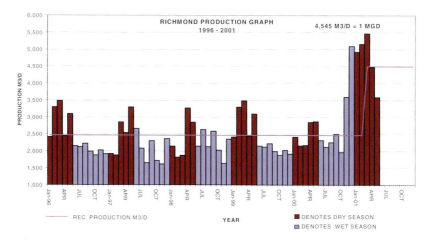

Figure 26. Water production trend plots of major plants over the last five years for the island of Tobago.

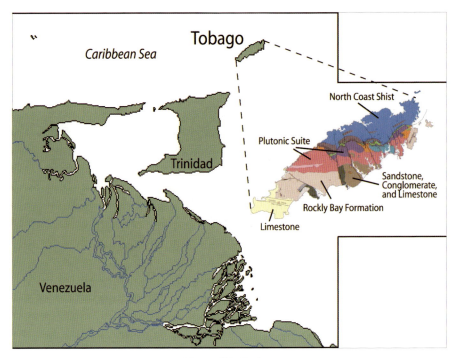

Figure 27. Setting.

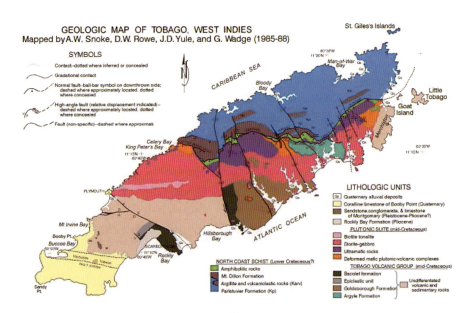

Figure 28. Geological map of Tobago.

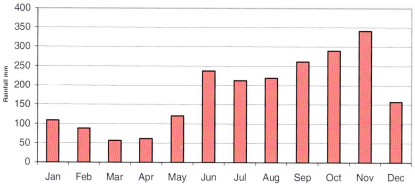

Figure 29. Ten-year average monthly precipitation (1988–1997).

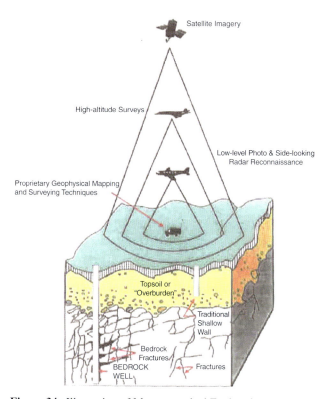

Figure 34. Illustration of Megawatershed Exploration Program.

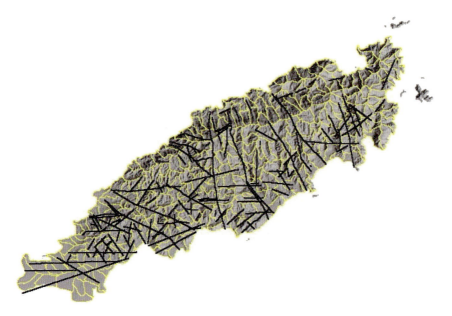

Figure 37. Implied fracture zones and polybasins.

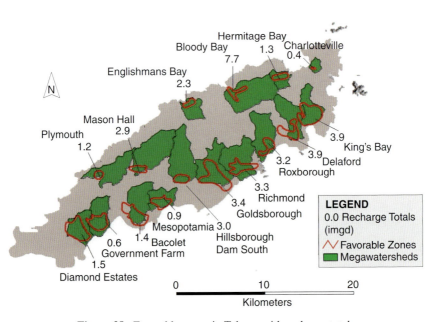

Figure 38. Favorable zones in Tobago with recharge totals.

Figure 43. ETI/HSA-Lennox Tobago Groundwater Exploration Program, favorable zones criteria matrix example.

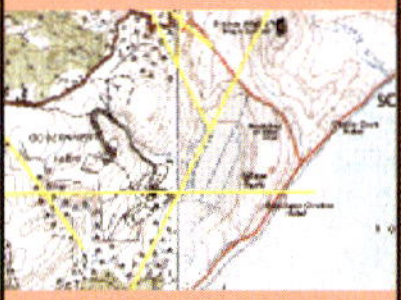

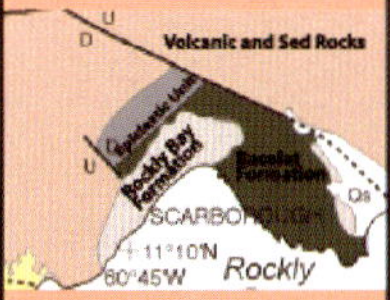

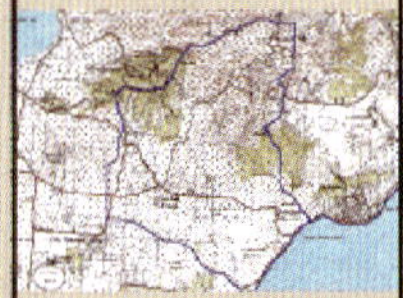

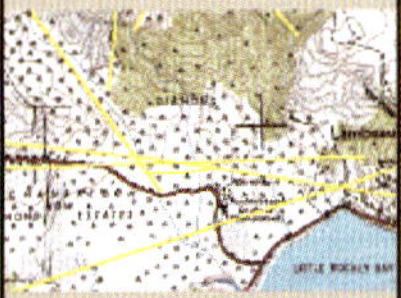

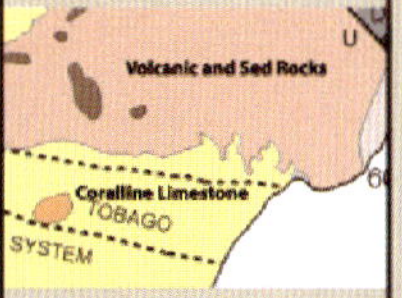

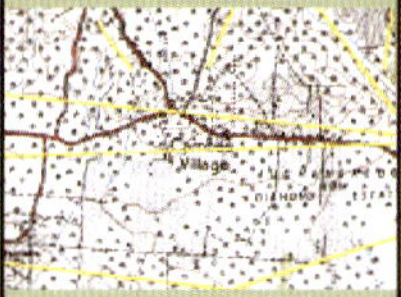

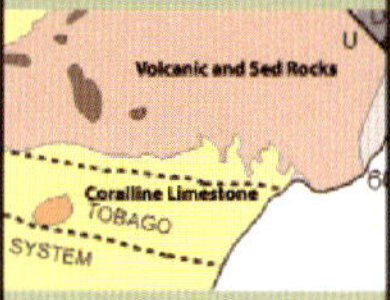

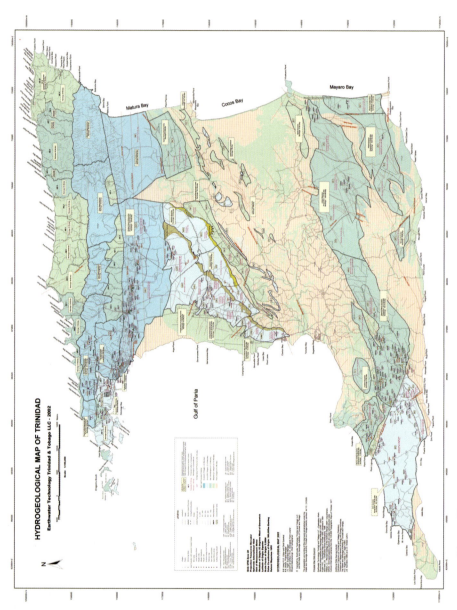

Figure 49. Hydrogeological map of Trinidad—2002 (Earthwater Technology Trinidad & Tobago LLC).

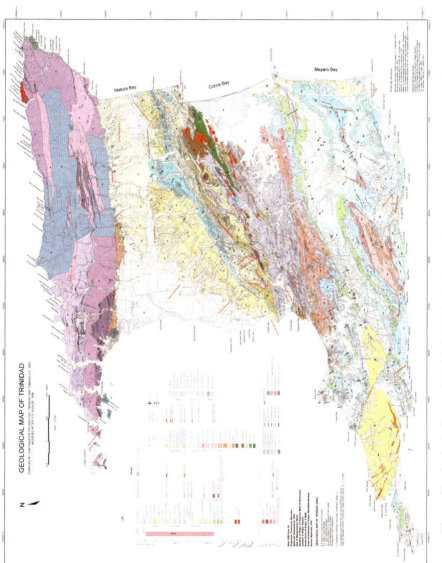

Figure 50. Geological map of Trinidad—2002 (Earthwater Technology Trinidad & Tobago LLC).

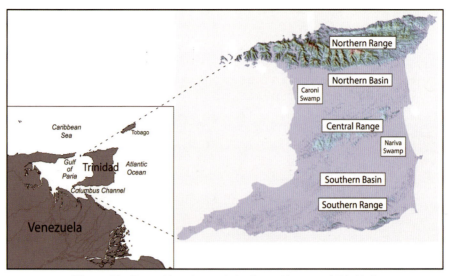

Figure 51. General location and physiographic map of Trinidad.

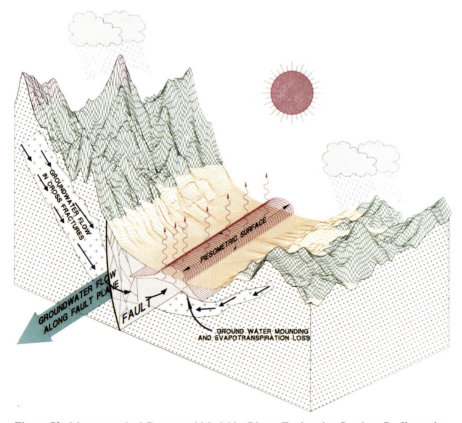

Figure 52. Megawatershed Conceptual Model by Bisson Exploration Services Co. Ilustration by Summit Engineering, 1992.

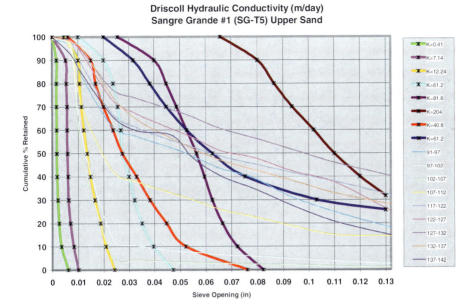

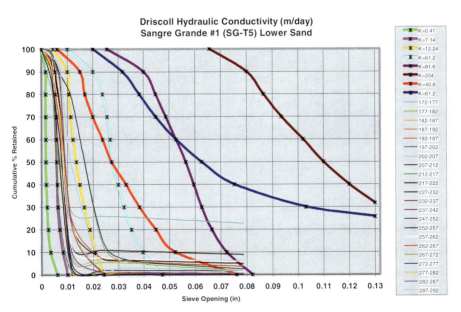

Figure 56. Size distribution curves from Sangre Grande #1 (SG-T5) plotted along with curves of known hydraulic conductivity [Driscoll, 1986]. Grain-size distribution curves of the upper sand (91′–142′). Grain-size distribution curves of the lower sand (172′–292′).

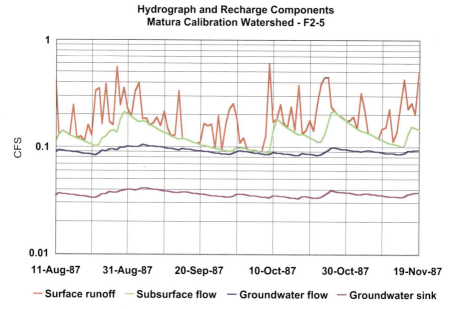

Figure 58. Hydrograph for storm events for the Matura watershed illustrate the hydrograph and recharge components.

Transmittable Storage	Potential Yield		
	m^3/Year	m^3/Day	gpm
Model 1	6.0×10^5	1,644	302
Model 2	6.05×10^5	1,650	304
Model 3	4.60×10^6	12,600	2,311

The most important aspect of the megawatershed model is that it predicts an order-of-magnitude higher groundwater flow at a site that fits the parameters of the model. Thus, it predicts the occurrence of high-grade water.

These model comparisons only serve as an example of the implications of the inclusion of fractured rock porosity and permeability in basin model calculations for safe yields. The procedure could be applied to any area that contains surface and subsurface drainage controlled by fault/fracture zones. The comparisons also support the original premise of the megawatershed model, which describes the boundary conditions related to the occurrence of high-grade water associated with a watershed-wide three-dimensional fracture network.

The exploration significance of the model is that it can be used as a pathfinder, to lead groundwater exploration teams to previously unsuspected targets that could yield large amounts of "high-grade" water.

Conclusions   The key features of the megawatershed model are as follows:

In arid coastal regions, bedrock-transmitted mountain precipitation, resulting from orographic effects, may considerably increase the amount of water available downgradient over what is available from local coastal plain recharge.

Zones of highly fractured bedrock terrain possess infiltration rates higher than the surrounding materials.

High fracture density and linear extent of these zones may contribute significantly to increased potential groundwater recharge.

Surface areas of alluvium with high infiltration rates may connect with highly permeable fracture/fault zones.

These fracture/fault zones may interconnect over large areas and thus provide a larger subsurface catchment area from which water flows, thus increasing the potential recharge. This may be represented as a surface expression of rectilinear drainage.

The fracture conduits containing high-grade water are restricted in lateral extent (width), are nonuniform, and comprise a small percentage of the total subsurface volume.

The envelope (and any hydraulically connected sediments) surrounding the fracture conduit serves as storage.

The best fracture/fault zones are those that have a high density of wide fractures parallel to a reasonable hydraulic gradient.

Highly favorable targets are those that contain geologic formations with primary porosity in hydraulic connection with fracture/fault zones and connected to the main source of recharge.

Exploration for, discovery of, and accurate mapping of these anomalous zones of groundwater flow are a tasks that require the combination of many disciplines: remote sensing, climatology, geomorphology, structural geology, hydrogeology, and geophysics.

Conceptual models, such as the megawatersheds model, help the exploration team to apply these disciplines in an efficient manner while increasing their probability of success.

REFERENCES

Abufila, T.M., 1984, *A Three-Dimensional Model to Evaluate the Water Resource of the Kufra and Sarir Basins, Libya.*

Ahmad, M.U., 1983, *A Quantitative Model to Predict a Safe Yield for Well Fields in Kufra and Sarir Basins, Libya.*

Alam, M., 1989, Water resources of the Middle East and North Africa with Particular Reference to Deep Artesian Groundwater Resources: Unpublished Report to The World Bank.

Angelillo, O.R., 1959, Replenishing Source of Waters Flowing Through Rock Fissure Aquifers: Unpublished Manuscript.

Badgely, P.C., 1965, Structural and Tectonic Principals, Harper and Row, New York.

Barbara, P.L., and Rosso, R., 1989, On the Fractal Dimension of Stream Networks.

Barker, J.A., 1988, A Generalized Radial Flow Model for Hydraulic Tests in Fractured Rock: Water Resour. Res.

Bisson, R.A., 1985–1989, Unpublished Research, and 1985, Confidential Report to BPSA, entitled: *Feasibility Study on the Development of Groundwater Resources in Southern Africa.*

Buckley, D.K., and Zeil, P., 1984, The Character of Fractured Rock Aquifers in Eastern Botswana: *Challenges in African Hydrology and Water Resources*

(Proceedings of the Harare Symposium), IAHS Publication No. 144, pp. 25–36.

Cederstrom, D.J., 1972, Evaluation of Yields of Wells in Consolidated Rocks, Virginia to Maine: U.S.G.S. Water Supply Paper 2021.

Cederstrom, D.J., Boswell, E.H., and Tarver, G.R., 1983, South Atlantic-Gulf Region; in Todd, D.K. (Ed.): *Ground-Water Resources of The United States*, Premier Press, California.

Chow, V.T., 1964, Handbook of Applied Hydrology: McGraw Hill, New York.

Cloos, H., 1948, The Ancient European Basement Blocks—Preliminary Note: Trans. Am. Geophys. Union 29, Vol. 1, pp. 1748–1759.

Dana, J.D., 1847, Origin of the Grand Outline Features of the Earth: Am. J. Sci., Sec. 2, Vol. 3, pp. 381–398.

deMarsily, G., 1985, Flow and Transport in Fractured Rocks: Connectivity and Sale Effect, in: Hydrology of Rocks of Low Permeability: Intern. Assn. Hydrologists, Vol. XVII, Pt. 1, pp. 267–277.

El-Baz, F., 1988, Origin and Evolution of the Desert: *Interdisciplinary Science Reviews*, Vol. 13, No. 4, pp. 331–347.

El-Baz, F., 1989, Monitoring Lake Nasser by Space Photography, *Remote Sensing and Large-Scale Global Processes*: Proceedings of the IAHS Third International Assembly, Baltimore, Maryland, IAHS Pub. No. 186, pp. 177–181.

Emery, J.M., and Cook, G.W., 1984, A Determination of the Nature of Recharge to a Bedrock Fracture System (results of testing a BCI Geonetics, Inc. production well at Putnam, Connecticut, U.S.A.): Paper presented at the National Water Well Association Groundwater Conference, Newton, MA, July 23–24.

Endo, H.K., and Witherspoon, P.A., 1985, Mechanical Transport and Porous Media Equivalence in Anisotropic Fracture Networks, in: Hydrology of Rocks of Low Permeability: Intern. Assn. of Hydrologists, Vol. XVII, Pt. 2, pp. 527–537.

Faust, C.R., Mercer, J.W., and Thomas, S.D., 1984, Quantitative Analysis of Existing Conditions and Production Strategies for the Baca Geothermal System, New Mexico: Water Resources Research, Vol. 20, pp. 601–618.

Gale, S.E., 1982, Assessing the Permeability Characteristics of Fractured Rock: G.S.A. Special Paper 189, pp. 163–181.

Henderson, G., 1960, Air-photo Lineaments in Mpanda Area, Western Province, Tanganyika, Arifa: A.A.P.G. Bull., Vol. 44, No. 1, pp. 53–71.

Hilgenberg, O.C., 1949, Die Bruchstructur der Sialischen Erdkruste: Akademie-Verlag, Berlin, Germany.

Hobbs, W.H., 1911, Repeating Patterns in the Relief and in the Structure of the Land: Geol. Soc. America Bull., Vol. 22, pp. 123–176.

Hopkins, W., 1835, *Researches in Physical Geology.*

Huntoon, P.W., and Lundy, D.A., 1979, Fracture-Controlled Groundwater Circulation and Well Siting in the Vicinity of Laramie, Wyoming: Groundwater, Vol. 17, pp. 463–469.

Issar, A., 1985, Fossil Water Under the Sinai-Negev Peninsula: Scientific America, Vol. 253, pp. 104–111.

Kohut, A.P., Foweraker, J.C., Johanson, D.A., Tradewell, E.H., and Hodge, W.S., 1984, Pumping Effects of Wells in Fracture Granitic Terrain: Groundwater, Vol. 21, pp. 564–572.

La Moreaux, P.E., Memon, B.A., and Hussein, I., 1985, Groundwater Development, Oasis, Western Desert of Egypt: A Long-Term Environmental Concern: Environ. Geol. Water. Sci., Vol. 7, pp. 129–149.

Lueder, D.R., and Simons, J.H., 1962, Crustal Fracture Patterns and Groundwater Movements: White Plains, N.Y., Geotechnics and Resources Inc. Report to Bureau of State Services, U.S. Dept. of Health, Educ. and Welfare, Final Report of Grant WP-53.

Mollard, J.D., 1957, Aerial Mosaics Reveal Fracture Patterns on Surface Materials in Southern Saskatchewan and Manitoba: Oil in Canada, pp. 26–50.

Moreno, L., Tsang, Y.W., Tsang, C.F., Hale, F.V., and Neretnieks, I., 1988, Flow and Tracer Transport in a Single Fracture: A Stochastic Model and its Relation to Field Observations: Water Resources Research, Vol. 24, pp. 2033–2048.

Nunn, K.R., Barker, and Bamford, D., 1983, In Situ Seismic and Electrical Measurements of Fracture Anisotropy in the Lincolnshire Chalk: Q.J. Engl Geol. London, Vol. 16, pp. 187–195.

Parizek, R.R., 1976, On the Nature and Significance of Fracture Traces and Lineaments in Carbonate and Other Terrains, in: Proceedings of Karst Hydrology and Water Resources Symposium, Dubrovnik, Yugoslavia, June 2–7, 1975, Water Resources Pub., pp. 47–108.

Price, M., 1985, Introducing Groundwater, Allen & Unwin, London.

Siddiqui, S.H., and Parizek, R.R., 1971, Variations in Well Yields and Controlling Hydrogeologic Factors, in: College of Earth and Mineral Science, Mineral Conservation Series Circular 82, Penn. St. Univ., pp. 87–95.

Smith, E.J., 1980, Spring Discharge in Relation to Rapid Fissure Flow: Groundwater, Vol. 17, pp. 346–350.

Sonderegger, J.L., 1970, Hydrogeology of Limestone Terraces—Photogeologic Investigations: Geol. Surv. of Alabama, NTIS PB-198.

Sterns, D.W., and Friedman, M., 1972, Reservoirs in Fractured Rock: AAPG Memoir #16, pp. 82–106.

Tsang, Y.W., and Tsang, C.F., 1987, Channel Model of Flow Through Fractured Media: Water Resources Research, Vol. 23, pp. 467–479.

Tsang, Y.W., and Witherspoon, P.A., 1985, Effects of Fracture Roughness on Fluid Flow Through a Single Deformable Fracture: in, Hydrology of Rocks of Low Permeability, Intern. Assn. of Hydrologist, Vol. XVII, Pt. 2, pp. 683–694.

1989, Tunnel Collapse Shows New Water Supply: *World Water*, Vol. 12, No. 4.

Zecharias, Y.B., and Brutsaert, W., 1988, Recession Characteristics of Ground-water Outflow and Base Flow From Mountainous Watersheds: Water Resources Res., Vol. 24, pp. 1651–1658.

5 Case Study—Tobago, West Indies 1999–2000
Unprecedented Groundwater Discoveries in a Small Island Developing State (SIDS)

UTAM MAHARAJ
Former Director of Water Resources Agency, Water and Sewerage
Authority of Trinidad and Tobago

INTRODUCTION

Space-age technologies incorporating images from satellites orbiting the Earth in a novel public–private sector partnership were used to discover and develop enormous quantities of high-quality groundwater for the island of Tobago. This is the first Caribbean application of these technologies and of this type of partnership, and Tobago's long history of conventional groundwater investigations, failed well drilling attempts, combined with a critical 2-mgd shortfall made it a "worst-case" island test venue for new groundwater discoveries.

Nevertheless, the use by the client of a "shared-risk" contracting approach, combined with the application by the contractor of the novel "megawatersheds" paradigm and an advanced state-of-the-art exploration program resulted in the identification of 66 MCM/Year (39.8 mgd) of previously undetected, renewable groundwater resources in the prevailing crystalline bedrock of the island. The contractor subsequently drilled and actually developed over 6.6 MCM/Year (4 mgd) of sustainable, spring-quality groundwater, all within one year. Under

Modern Groundwater Exploration: Discovering New Water Resources in Consolidated Rocks Using Innovative Hydrogeologic Concepts, Exploration, Drilling, Aquifer Testing, and Management Methods, by Robert A. Bisson and Jay H. Lehr
ISBN 0-471-06460-2   Copyright © 2004 John Wiley & Sons, Inc.

the contract, the contractor was only paid for water actually developed (and tested) and warranteed the wells for one year.

Tobago's present and future water problems were resolved in a single year's work, using best-available technology, at minimal risk to the client. This project has proven that with proper risk-sharing arrangements, the best expertise and provider of technology for groundwater exploration and development can be acquired to develop new sources of groundwater in difficult geological terrains not previously thought to be available.

WATER DEVELOPMENT TEAM

The contracting team was led by Earthwater Technology International, Inc. (ETI), which designed and supervised the project in joint venture with a Trinidad water and oil drilling firm, Lennox Petroleum Services Ltd., which also accepted the risk of financing this unprecedented Caribbean exploration project. The technical team was directed by ETI's Robert A. Bisson and included core members of his original BCI-Geonetics Africa team, Dr. Roland B. Hoag, Jr. and Joseph Ingari, together with a highly motivated group of field investigators under geologist David Hisz and fast-learning drillers using new technologies under the leadership of Lennox Persad and Carlyle Charles. Consulting geologist David Rowe's field mapping skills and prior published work on Tobago's geology was a valuable addition to the team, and the client's team was directed by Water Resources Agency Director Utam Maharaj with able support from Tobago Water Supply manager Oswyn Edmund.

BACKGROUND

The World Bank and United Nations report that over much of the globe, human demand will exceed known sources of natural fresh water supply early in this century.[1] It was estimated that currently one-tenth of the world's population does not have access to adequate potable water, and by the year 2025, this proportion will increase to one-third.[2] This is equivalent to 2.8 billion people. Solutions to the water supply-demand situation require new ways of doing business using new technologies to solve long-standing problems. The island of Tobago is one of the smaller islands within the Caribbean. The island is blessed with beautiful beaches and tropical climate, and this strategic location was faced with the prospect of public health threats and stunted economic

Total Demand=13.3

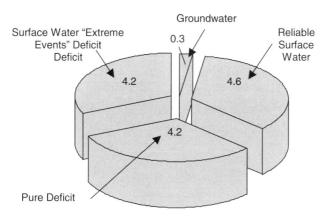

Figure 25. Surface water, groundwater, and deficit before project. Units are in MCM/Year.

development due to a severe supply-demand water deficit that was progressively worsening. The economy of this pristine island is almost entirely dependent on tourism.

In the year 2000, the island, with a population of 55,000 and an enhanced seasonal demand due to tourism, had a computed water supply demand deficit of 3.3–5.0 MCM/Year (2.0–3.0 mgd) or approximately 35%. Figure 25 shows the water supply-demand situation on the island. The problem of water supply was further exacerbated by the seasonal and spatial water availability on the island. Extreme events on the island due to its tropical weather pattern resulted almost annually in drastic cuts in potable water supply during both the dry season and the wet season. The dry season water stresses were due to the river lowflows, reflecting the 97% dependence of the island's water supply on surface water sources, whereas the paradoxical rainy season stresses were due to natural steep terrain and enhanced development on the island associated with road and building construction and quarrying. This resulted in high silt loadings, which the current intakes and treatment facilities could not handle. Often they had to be turned off during heavy rainfall, and water was severely rationed to domestic and tourist customers. During these dry and wet season periods, the island's average "theoretical" water supply-demand deficit would virtually double. The estimated increased deficit due to extreme events was 4.2 MCM/Year (2.5 mgd).

In 2000, a year with average precipitation, few domestic customers received regular supplies, and it was not uncommon in both the dry and wet seasons for hotels, which possessed internal storage systems designed to supply their patrons with a 24-hour supply, to constrain usage to 4 hours per day. In fact, the opening of a five-star luxury hotel was delayed for over one year at enormous cost to the developer, and the $100 million hotel was among the first beneficiaries of the high-yield megawatersheds wells.

Figure 26 shows the production profile for the three major plants in Tobago over a five-year period prior to groundwater development. Hillsborough reservoir shows an extreme dry-season vulnerability due

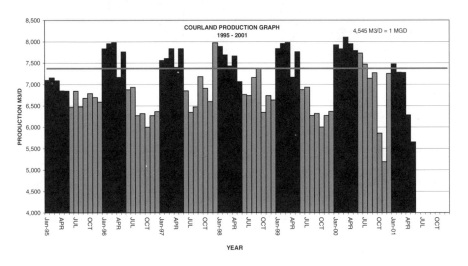

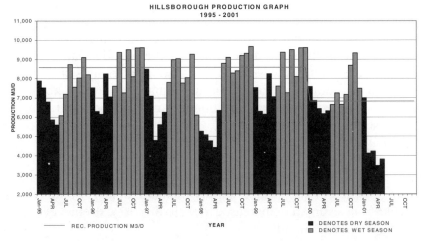

Figure 26. Water production trend plots of major plants over the last five years for the island of Tobago. (See color insert.)

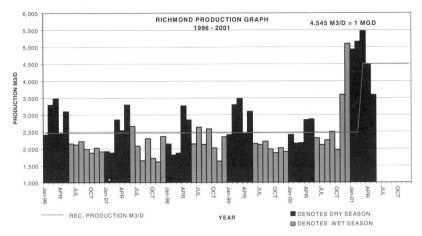

Figure 26. *Continued*

to overabstraction from the reservoir every year. The darker bars show monthly abstractions averaged daily. For this reservoir during the rainy season, the reservoir delivers close to its design capacity shown by the pink line; however, during the dry seasons, severe rationing must occur. On the other, Courland and Richmond plants traditionally demonstrate wet season susceptibilities due to the large influx of silt from their watersheds.

Even in an average rainfall year, domestic customers received water supplies for a few hours a day and for a few days a week. It was not uncommon in the height of the tourist seasons for hotels, which possessed internal storage systems designed to supply their patrons with a 24-hour supply, to constrain usage to 4 hours per day.

The island was in the midst of water crises that, if not abated, would cause unprecedented hardship on the population, discourage foreign investment, and derail economic growth. Several options were being considered to cure the problem, including the building of a new impounding reservoir, seawater desalination, water barging, upgrading the water treatment plants, and groundwater development.

The Government of Tobago was considering several options to cure the problem, including the building of a new impounding reservoir, seawater desalination, water barging, and upgrading the water treatment plants and groundwater development. The first options were considered very costly or environmentally unacceptable, whereas the groundwater option was considered to represent a very high risk of failure, because expert investigations by both local and foreign geologists, over a period of 40 years, had concluded that Tobago possessed a total of less than 0.83 MCM/Year (0.5 mgd) of proven supply (all of it already used), and no further potential for groundwater existed on the island.

Prior to the year 2000, the Water and Sewerage had been exploring the island since 1958 and had drilled ten water wells. Dismal results had been obtained. Virtually all of the wells are poor producers or produce water of poor quality by present standards. The result of 42 years of exploration had delivered three useable production wells, which were kept in operation. These had a combined production of 0.2 million gallons per day or less than 3% of the 5.6 million gallons per day existing production capacity. Almost all wells were distributed throughout the southwestern portion of the island where the major water demand but lowest rainfall existed. The wells were located in the Cenozoic age coralline sedimentary deposits and alluvials sediments. Two production wells located at Government Farm had a combined potential of 100,000 gallons per day, and the single well in the north central portion of the island was completed in alluvial gravels. Four of the other alluvial and three of the coral wells were nonpotable.

The first options were considered very costly or environmentally unacceptable, and the groundwater exploration option was considered to represent a very high risk of failure, because experts both local and foreign, as recently as in a 1999 report had concluded that Tobago possessed a total of less than 0.83 MCM/Year (0.5 mgd) of proven supply (all of it already used), and no further potential for groundwater existed on the island.[3] The favored solution was the construction of a new impounding reservoir that would store water in the rainy season and deliver 8.3 MCM/Year (5.0 mgd) during the dry season. This option was estimated to take eight years to materialize with a capital cost in the region of $60 million and producing substantial environmental impact on the surrounding ancient, protected rainforest.

In the late 1990s, the client, Water & Sewerage Authority of Trinidad and Tobago (WASA), facing the financially daunting specter of a major public works project, was made aware of new technologies and geological concepts successfully used elsewhere to locate and develop groundwater within complex geological environments similar to Tobago. As far as WASA was concerned, there was no more groundwater in close proximity to the demand that could be developed, because all known aquifers were producing at their safe yields. In the absence of prior experience and expertise within WASA in this new type of groundwater development, embarking on such a project would be an extremely high risk.

An internal evaluation of the concept of risk and of how best to manage risks led to WASA initiating a means of identifying and engaging a suitable contractor to undertake an integrated groundwater exploration and development project.

SETTING

Tobago is the smaller of the two-island Republic of Trinidad and Tobago. The Republic is the southern most of the chain of Caribbean island countries of the Lesser Antilles (Figure 27). Trinidad is located just north of Venezuela's coastline, and Tobago lies 22 miles northeast of Trinidad. The island physiography is characterized by a southwest to northeast trending mountain ridge for over two-thirds of the eastern part and the minor part in the western portion characterized by a low-lying flatlands. Its size is $300\,km^2$.

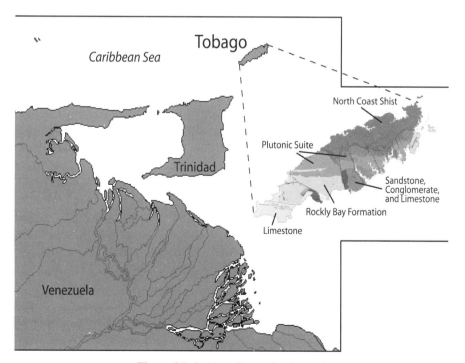

Figure 27. Setting. (See color insert.)

LOGISTICAL CONSTRAINTS AND CHALLENGES

In 2000, the island of Tobago was largely undeveloped, with a limited delivery network for water and electricity, few roads, especially into the interior, and virtually no indigenous heavy industry to support construction work. Most water mains were old, small in diameter, and in poor condition. Due to WASA budgetary limits, prospective drilling targets were kept close to existing water mains, and less than two kilometers of new pipe and electrical wire was allowed to connect all of the new production wells.

In addition, the U.S.-based Earthwater Technology team specified drilling technologies for Tobago's volcanic bedrock that mandated ETI's local drilling partner Lennox Petroleum purchase and import all new equipment from Canada and the United States. In fact, all drilling and construction materials were transported by ship to the island from Trinidad or North America, via Trinidad, a time-consuming and costly process.

The northern two-thirds of Tobago, where most of the geological fieldwork was planned, comprised heavy bush and rainforest, where access was limited to footpaths and helicopters. However, these challenges were counterbalanced by the stunning physical beauty of Tobago and its natural setting, the island's splend genial residents, and colorful fauna and flora.

GEOLOGY AND HYDROLOGY

The island is an exposure that forms part of a structural high comprising chiefly Mesozoic igneous and metamorphic rocks (see Figure 28). Tobago resides at the northeastern most boundary of the present-day South American continental shelf. Recent investigators [Wedge and MacDonald, 1985; Frost and Snoke, 1989, Figure 3] have posited that Tobago, along with the Villa de Cure klippe, the Netherlands Antilles, and various Venezuelan offshore islands (e.g., La Blanquilla) is a member of a composite Mesozoic oceanic-arc terrane gradually accreted to northern South America, perhaps beginning as early as Mid–Late Cretaceous.

The Mesozoic rocks of Tobago can be divided into four main groups, which form approximately east–west-striking belts that transect the island. They are the North Coast Schist, amphibolite-facies aureole, the

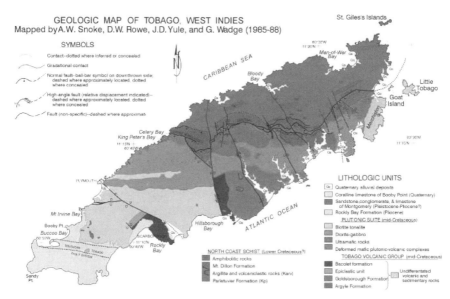

Figure 28. Geological map of Tobago. (See color insert.)

ultramafic to plutonic complex, and the Tobago Volcanic Group. Brittle faults of variable slip delineate several fault systems that have fragmented the Mesozoic belts of the island. In addition to the large-scale fault systems with associated offsetting lithologies, the island is pervasively fractured by faults of all scales. Younger Cenozoic sedimentary deposits consist of clays, silts, sandstones, and gravels (Rockly Bay Formation), and an even younger platform of coralline limestone (Bobby Point).

The Mesozoic igneous and metamorphic rocks of Tobago constitute a cross section through a fragment of an allochthonous, composite oceanic-arc terrane. New geologic mapping coupled with petrochemical and geochronometric studies indicate the following:

The protolith for the oldest exposed part of the oceanic-arc terrane (NCS) is pre-Aptian and perhaps as old as Late Jurassic. The NCS experienced pre-Albian penetrative deformation and lower greenschist facies metamorphism. The Tobago plutonic complex and Tobago Volcanic Group are cogenetic and evolved in the mid-Albian. Ultramafic rocks of the plutonic suite intruded and dynamothermally metamorphosed the NCS forming an inverted, amphibolite-facies metamorphic aureole. A mafic dike swarm widely intruded into the volcanic-plutonic complex and locally intruded the amphibolite-facies aureole and NCS. The mafic dikes are chiefly mid-Albian and therefore were clearly coeval with the volcanic-plutonic suite, although a member of the swarm has also yielded an igneous crystallization age as young as 91 Ma. The Mesozoic lithic belts of Tobago have been attenuated and offset by brittle fault systems that were related both to the late-stage emplacement history of the mid-Cretaceous plutonic complex as well as much later Cenozoic strike-slip faulting. The Cenozoic deposits on Tobago represent various shallow marine accumulations and their preservation therefore records tectonic uplift or sea level change.

The island's climate is characterized by typical tropical weather patterns, with a dry season spanning from January to May and a rainy season from June to December. The 10-year monthly average rainfall at the center of the island (at Hillsborough Dam) is shown in Figure 29. Temperatures range from 25°C to 27°C, and humidity ranges between 50% and 100%.

Average rainfall ranges from 3800 mm on the main ridge to less than 1250 mm in the southwestern lowlands. The isohyetal developed for the island by the local water authority using 30 years of data collected at 24 stations, albeit with several gaps, gave a fairly accurate isohyetal as shown in Figure 30. Stream flow data were routinely collected at five representative watersheds on the island since 1973. Evaporation and other related data have been collected at a single location on the southwest of the island (at Crown Point) since 1971.

Prior to the megawatersheds exploration program, no island-wide, comprehensive groundwater investigations were conducted on the Island of Tobago, and all efforts at developing groundwater have been

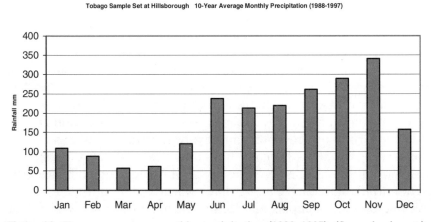

Figure 29. Ten-year average monthly precipitation (1988–1997). (See color insert.)

Figure 30. Thirty-year isoheyetal map.

piecemeal evaluations, principally in sand and gravel areas. Review of previous reports revealed largely unsuccessful attempts to discover and develop groundwater in Tobago. This reflected the limitations of traditional hydrogeological concepts focused on the presence of groundwater in sand and gravel aquifers. The potential of the island's traditional alluvial aquifers and their locations is shown in Figure 31. The cumulative potential of these aquifers indicates a groundwater resource of 1.0–1.5 mgd.

Unfortunately, the failure of past efforts led investigators to the false conclusion that Trinidad and Tobago lack abundant groundwater resources, impeding the water authority to solve the acute water shortages on the island.

Most of Tobago's present-day coastal communities specifically in the southwest of the island, where the fresh water needs are greatest, are located on land that was below sea level 10,000 years ago, very recently in geologic time. These once-submerged lands retain salt water within

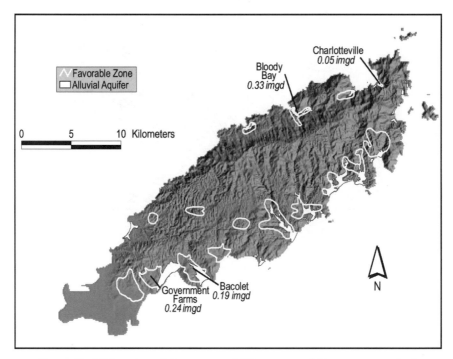

Figure 31. Highest benefit/cost target drilling zones within megawatershed.

the pore spaces of their sediments, which may be left intact, diluted, or even replaced by rain-fed fresh waters, depending on the vagaries of local geologic structure and climate. Consequently, wells drilled blindly into these sediments have yielded brackish water. Several other wells drilled into alluvial deposits gave low-yielding wells at Bloody Bay and Government Farm. The history of documented groundwater development efforts in Tobago is summarized in Figure 32.

As with many other areas of the world, conventional water balance concepts and calculations have little in common with the actual fresh water resource available to consumers. In fact, the overall theoretical fresh water availability in Tobago from rainfall was computed to be 141 MCM/Year, which, if expressed on a per capita basis, yields a value of 2560 CM per year. This value, if compared with the World Bank's standards, indicates that the island cannot be technically classified as a water-scarce country, with its per capita demand of 236 CM/Year.

It is interesting to note that the megawatersheds paradigm and exploration technology verified through drilling an island-wide ground-

DATE DRILLED	AREA	GEOLOGIC ENVIRONMENT	# OF WELLS	RESULTS	STATUS
1911/1912	LOWLAND/COVE	TERTIARY	2 For Oil	No Oil	–
1958	LOWLAND ESTATES	TERTIARY	1	23m³/d	Abandoned
1965	MT. IRVINE	CORAL	10 Borings for Dam Site Foundation	–	–
1974	GOLDEN GROVE	CORAL	Evaluated Existing Dug Well	Brackish	Abandoned
1982	BON ACCORD	CORAL	2	Brackish	Abandoned
1982/1989	GOV'T FARM	LIMESTONE	2	274m³/d	In Use
1983	GOLDEN GROVE	CORAL	1	Brackish	No Production
1990	BLOODY BAY	ALLUVIALS	1	574m³/d Fresh	In Use
1990/1991	COURLAND	ALLUVIALS	2	Brackish	Abandoned
1990	MT. IRVINE	ALLUVIALS	1	No Water	No Production
1991	PALM TREE	ALLUVIALS / CORAL	1	22m³/d/ elevated H₂S	In Use
		TOTALS	23		870 m3/d (0.19 mgd)

Figure 32. History of groundwater development on Tobago.

water resource of approximately 66 MCM/Year, or 1200 CM/year per capita, which can actually be readily and economically accessed by the population.

TECHNICAL PROGRAM

The Year 2000 project targeted high-yield wells in the island's under-lying bedrock, comprising igneous and metamorphic rocks, as new sources of freshwater for the island. The contract required the ETI-Lennox team to reassess the hydrogeology of the island using their novel, megawatershed hydrogeological conceptual model, identify favorable locations proximal to WASA's distribution system, and test-drill and build production facilities to deliver a minimum of 2 mgd (over 4 mgd was actually drilled and tested) to WASA's mains. This required that the team accurately determine the full groundwater recharge, storage, and transmission potential of the island and each target zone and verify the new megawatersheds models through short- and long-duration pumping tests.

MEGAWATERSHED PARADIGM

The megawatershed paradigm, or model, is a modern, geology-based concept of groundwater environments, which has ancient roots, but could not be adequately understood and substantiated using "the scientific method" until the availability of space-age technologies. These technologies include Earth observation satellites and geographic information systems (GIS), combined with the advent of modern geological concepts, especially plate tectonics, enabling hydrogeologists to objectively re-evaluate the Earth's water resources.

Documentation of groundwater occurrence in deep bedrock mines and wells extends thousands of years into history. Examples of this phenomenon include deep groundwater infiltration forcing the closing of ancient Egypt's hand-dug gold mines [Page, 1983] to the constant dewatering required in present-day South Africa's kilometers-deep shafts and Nevada's open pit excavations [Stone et al., 1991]. In modern times, petroleum geologists and other deep well drillers have documented fresh water discoveries at depths exceeding 3 km in fractured rocks [Magaritz et al., 1990; Burbey and Prudie, 1991; Huntoon, 1986; Moore, 1996; Simmons, 1992]. As little was known about the regional geometry, hydraulic conductivity, and connectivity of bedrock fault and fracture zones, and the paucity of worldwide rainfall and evapotranspiration data, hydrologists conservatively concluded that deep groundwater sources were principally "fossil" in nature, disconnected from active surface recharge, and of unknown, ancient origin [USGS, 1998; Gleick, 1993]. These assumptions have historically been applied to water resource evaluations of the world's great groundwater catchments, from Egypt's immensely thick sandstone and carbonate basins to the equally vast Great Carbonate Basin of the western United States.

Since the 1970's and 1980's, advent of modern Earth observation satellites and general acceptance by the world community of Earth scientists of plate tectonics as valid scientific theory, geologists have commenced using tectonic models and satellite imagery to map tectonically induced crustal fracturing for a variety of practical purposes, including economic recovery of oil, gas, minerals, and, more recently, deep groundwater. Modern investigators have found evidence of many types of tectonically related deep groundwater occurrence, from paleowaters slowly flushed through the Earth's porous and fractured crust by continental-scale tectonics over millions of years [Person et al., 1996] to active introduction of modern meteoric waters in the upper 1 km of fractured bedrock [USDOE, 1998]. Other scientific papers and reports added credence to the thesis that deep, regional, interbasin, and coastal groundwater flow systems demonstrated active regional recharge, but did not address specific origins, pathways, or seek to quantify potential recharge [USGS, 1991; Driscoll, 1997; Church, 1996; Falkowska, 1998; Mirecki and Manheim, 1998].

In the mid-1980s, investigators working in East Africa for USAID first correlated regional tectonic frameworks with hydraulically conductive fracture geometries and groundwater occurrence and subsequently performed detailed studies of Northwest Somalia and the coastal region of Sudan that addressed specific attributes of interbasin, fractured bedrock groundwater flow systems such as fracture geometries, conductivities, catchment, and recharge [Bisson et al., 1986, 1987, 1989]. The term "megawatershed" was formally introduced into the lexicon of hydrogeological terms in 1989 and 1990 publications describing the East Africa Rift-related groundwater phenomena [Bisson and El-Baz, 1990] (Figure 33). The original investigators' integrated, systematic approach to megawatersheds exploration became known as the "Megawatersheds Exploration Program" (Figure 34). However, until the Tobago Project was launched, there were no known existing case studies involving complete hydrogeological systems that quantitatively compare results from actual applications of the "megawatersheds" paradigm of deep groundwater occurrence and the traditional basin and "watershed" concept of aquifers.

MEGAWATERSHED EXPLORATION

For over 25 years, ETI explorationists evolved and successfully employed a state-of-the-art program of groundwater exploration adapted from modern geological concepts and technologies and interpretive methods commonly used in modern oil, gas, and mineral exploration but rarely employed in a similar fashion by hydrologists. The

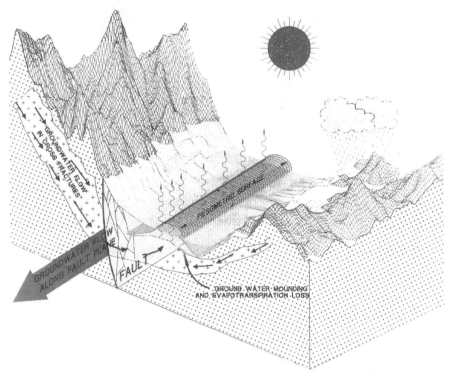

Figure 33. Conceptual megawatershed model (created by Bisson Exploration Services and Summit Engineering, 1993).

"megawatershed" paradigm of groundwater occurrence is the template upon which the exploration program is constructed (Figure 34).

The types of data used in the megawatersheds exploration program for this project included geological, hydrological, topographic, cadastral, geophysical, and remotely sensed data, along with existing GIS databases of cultural information, which are summarized in Figure 35.

GIS integration of the diverse remotely sensed data and previous geological interpretation of the island was used as the base for overlying diverse data sets, allowing for superior definition of the fracture fabric system of the island. Digital topographic data for Tobago acquired included both a digital raster image of the 1:25,000 topographic map and a digital elevation model (DEM) with a 30-m resolution. All imagery was also combined with DEMs to provide a third dimension to the interpretation process. This integration allowed data manipulation by changing vertical exaggeration and viewing of the imagery from many different angles by changing the sun's angle and

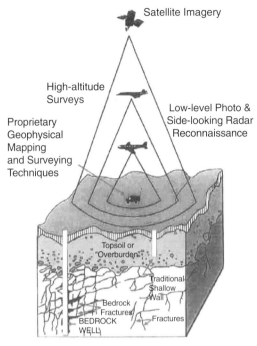

Figure 34. Illustration of Megawatershed Exploration Program. (See color insert.)

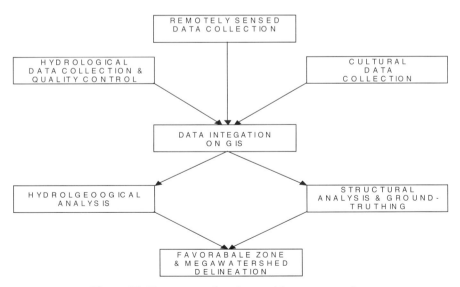

Figure 35. Remote-sensing data and image processing.

azimuth, thereby giving several three-dimensional perspectives to the various sets of imagery. A topographic image in raster format also integrated into the GIS was used to identify the precise locations of various geologic features on the ground.

The hydrological assessment of the island used existing rainfall data and correlated data overlain on the microcatchments of the island to apply the water balance equation for the hydrological analysis of this project.

Other GIS files obtained from the local water authority included the water distribution system and location of water withdrawal points, including wells and dams. These data in conjunction with the location of demand centers were used to assess the most cost-effective locations for well siting and avoidance of groundwater contamination. Copies of the cadastral maps for Tobago were acquired from the Trinidad and Tobago Department of Lands and Surveys. Paper copies of 1:25,000 and 1:10,000 scale topographic maps were acquired from the Trinidad and Tobago Department of Lands and Surveys.

REMOTE-SENSING DATA AND IMAGE PROCESSING

The team acquired Thematic Mapper (TM) Imagery acquired from the LandSat satellite, which had a resolution of 30 m for the reflectance spectra, with the exception of the thermal infrared, which had a resolution of 120 m. The image that had minimal cloud cover that was used in this project was a LandSat V TM image, taken on December 14, 1985. Team technologists digitally processed the images, geocoded the resulting data files to the Tobago 1:25,000 topographic maps, and incorporated them onto the GIS database.

Further processing by geologists of a three-band image combination gave information on the morphology and structural features of the island's surface. This band combination in low-latitude, tropical environments penetrates tropical haze better due to the use of longer infrared TM bands. This image is shown in Figure 36. A six-band processing technique (GeoView), using a proprietary ratioing technique, then allowed the characterizing of the lithologic units in areas of limited vegetative cover.

The thermal infrared band is used to differentiate surface temperatures, with cooler temperatures showing up as darker colors. Water-filled fractures and saturated alluvial and groundwater discharges to

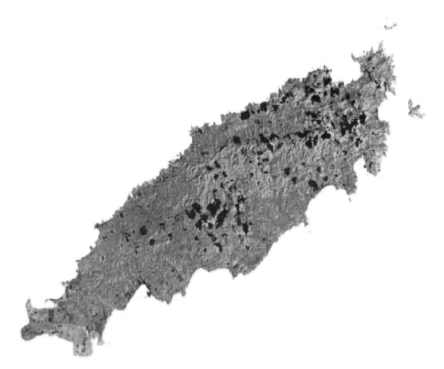

Figure 36. Landsat Thermatic Mapper™ image of Tobago.

the sea may be discernable with the thermal infrared. However, in a tropical environment such as Tobago, where the soil is moist most of the year, there is much less contrast between saturated and unsaturated soils. For Tobago, the thermal infrared image was of some, albeit limited, use.

Satellite radar imagery (taken on June 14, 1999 of resolution of 8m) that penetrates vegetation was used to discern topographic features and surface roughness to differentiate rock types.

Aerial Magnetic Surveys sourced from the Trinidad and Tobago Ministry of Energy also aided determination of the mineralogy of rock types for lithological characterization.

Aerial photography was acquired from the Trinidad and Tobago Department of Lands and Surveys. The photography acquired was at a scale of 1:25,000 and was flown in 1994. These photograph yielded 3-D views of the island's surface to infer the presence of lineaments and fractures and refine the satellite and DEM-derived megawatersheds models.

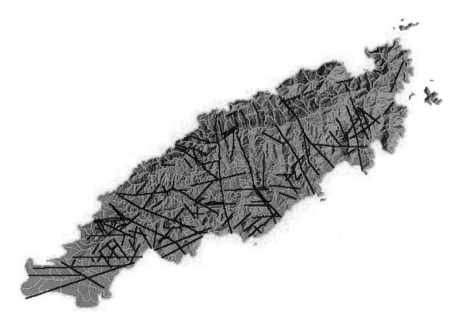

Figure 37. Implied fracture zones and polybasins. (See color insert.)

Expert interpretation of these multiple data sets in addition to an in-depth understanding of the regional and local tectonics is the key to determining the presence of bedrock aquifers, the extent of the megawatershed, and the exact location of the target aquifers and recharge zones into the aquifers. The reinterpreted structural geology of the probable open fracture networks of the island with ground-water recharge units (GRU or polybasins) is shown in Figure 37. GRU is discussed further on in this case study.

The hydrology of these open fracture systems and associated poly-basins were analyzed and amalgamated into megawatersheds and then subdivided into favorable zones representing best drilling sites near existing water mains, with estimated recharge totals (Figure 38).

Shaded-relief images of Tobago were generated using DEMs created from 1:25,000 scale topographic maps of Tobago with 30-m sample spacing. Several shaded-relief images were then generated using the Golden Software Surfer program (Figure 39). These programs allow the exploration team to select the specific light position angles for the shading, including azimuth and vertical position. The amount of vertical exaggeration and color/contrast qualities can also be varied. The DEM was also applied to the various images generated. For this study, many different shaded-relief images were generated from a variety of

Favorable Zones in Tobago with Recharge Totals

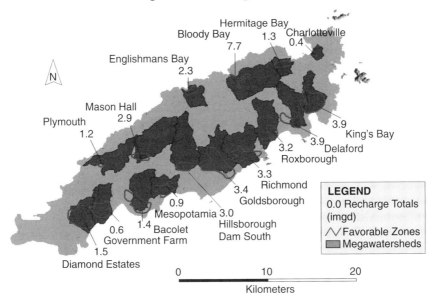

Figure 38. Favorable zones in Tobago with recharge totals. (See color insert.)

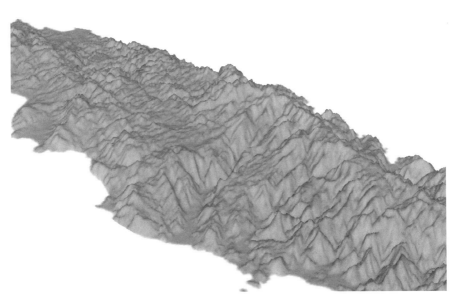

Figure 39. Digital elevation model of Tobago.

light position angles. Geologically analyzed linear features, called "inferred fracture zones," were highlighted on each of the images for each light position, and a compilation of all features generated from the five images was prepared as a data layer in ArcView for integration with the other data sets.

GEOLOGICAL ANALYSIS

Structural analysis of the DEM inferred fracture zones identified on Tobago in context with field mapping and geotectonic models suggests that widespread brittle deformation has affected the southwestern half of the island. The fact that most of these linear features do not offset mappable lithologic boundaries suggests that they indicate limited displacement structures. However, the inferred fracture zones also contain two features that extend for more than 15 km in an east–northeast direction (Figure 40).

One of these inferred fracture zones includes a prominent north-facing Holocene fault scarp in the Courland Bay area mapped by the

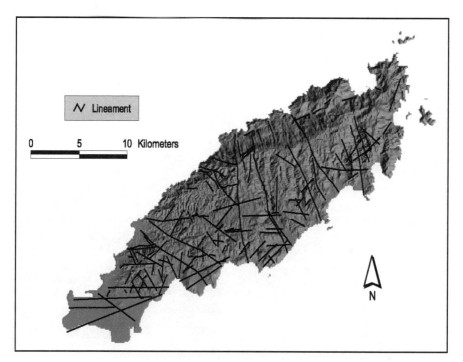

Figure 40. Inferred fracture zones.

field team, suggesting that this fault is part of the Southern Tobago Fault System. A similar structure extends from the Crown Point area through the north end of the Bacolet Bay Quaternary basin.

In addition to the east–northeast linear feature set, northwest-trending features and less frequent northeast-trending linear features were also observed. Most of the older mapped northwest-trending faults do not show up as discrete linear features because they occupy fairly broad valleys. These valleys were probably formed over geologic time by the differential erosion of the less-resistant fault zones.

A structural pattern analysis of the GEOPIC image confirmed linear features similar to those shown on the DEM data and showed several northwest-trending structures in the eastern part of the island not seen on the DEM image. Several of these northwest-trending features were subsequently confirmed as fracture zones and mapped faults.

The radar image was of limited use, corroborating the extensive east–northeast striking linear feature identified on the. In addition, several northeast and northwest linear features were identified.

These east-northeast-trending strike-slip fractures were verified by field inspection as being the youngest fractures. The linear structures that were mapped from the DEM, radar, and satellite images were sub-divided based on azimuth and length.

The original working copy of Arthur W. Snoke's Geologic Map of Tobago [1998] (Figure 27) was digitized and digitally "draped" over the GeoViewTM satellite image using ArcView. The purpose of overlaying the GeoViewTM image with the geologic map was to refine the mapped lithologic boundaries. Results of brittle fracture analyses of the DEM, radar imagery, and GeoPic image were input into the GIS program as separate layers.

Using image-maps resulting from the remote-sensing and geological map analyses, the ETI team carried out extensive field studies of Tobago's geology to verify their preliminary conceptual models of the island's tectonic history and brittle fracture geography.

HYDROLOGICAL MODELING

Hydrological modeling was employed to determine the amount of groundwater infiltration into the extensive fracture network of the newly delineated megawatersheds. Ideally, it is desirable to have at least ten years of continuous stream flow and precipitation data to model daily infiltration rates of groundwater and obtain a good estimate of aquifer recharge. It is also advantageous to have evaporation data at several points reflecting the variable topography of the island for calculation of evapotranspiration and water balance.

These data were not available for Tobago, and an alternative approach was designed to determine groundwater infiltration rates into megawatersheds. Instead, climatological and gauged stream flow data recorded by WASA over many prior years was used in this investigation to estimate groundwater recharge to both bedrock and alluvial aquifers. Categories of data analyzed include the following:

- Rainfall isohyetal produced by WASA from 30 years of data.
- Annual stream flow data from the Courland gauged station for the years 1977, 1981, 1985, 1986, 1987, and 1990.
- Annual stream flow data from the Louis D'or gauged station for the years 1984, 1985, 1986, 1987, 1988, 1989, and 1990.
- Annual average basin-wide precipitation data were compiled by WASA for the Courland, Louis D'or, Kings Bay, and Richmond watersheds for the same years as the annual flow data.
- Annual stream flow data from the Kings Bay station for 1976 and the Richmond station for 1987.
- Free surface evapotranspiration data obtained from WASA files for the Crown Point airport station in Tobago.

RECHARGE CALCULATIONS

Critical to the computation of the safe-yield of an aquifer or water well is the determination of the groundwater recharge computed from the equation:

$$\text{Recharge} = \text{Precipitation} - \text{Evapotranspiration} - \text{Run-Off}$$

As described by Hoag et al. [2000], it was necessary, due to limited data, to develop relationships between the hydrological parameters for which data were available. A consistent proportional relationship was established between precipitation and evapotranspiration at a single location and between precipitation and runoff for data at several watersheds on the island.

ETI applied this relationships to the 30-year isohyetal map and created a detailed groundwater recharge map, consisting of 240 1 km^2 sub-basins, called Groundwater Recharge Units (GRUs), from which groundwater recharge was computed for megawatersheds and local

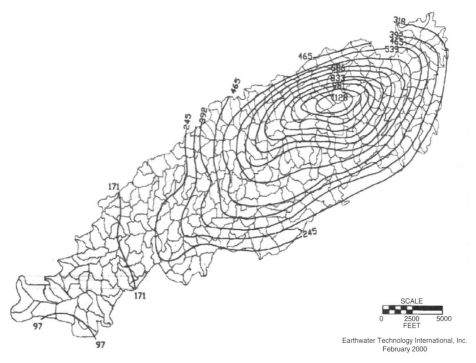

Figure 41. Recharge contours and groundwater recharge units (GRU).

aquifers by composite analysis using ArcView. ETI then created the groundwater recharge map of Tobago (Figure 41) using a proprietary software package.

The megawatershed recharge model combined the mapped mega-watersheds with the distribution of GRUs. Where inferred fracture zones cross major topographic watershed boundaries, the GRUs associated with these fractures are summed to estimate the recharge to the megawatersheds and fracture systems associated with drilling targets in favorable zones, as described below. The result of this analysis is shown in Figure 42 favorable zones with production wells are labeled. Table 9 illustrates the substantial difference between the traditional and megawatershed groundwater recharge models of Tobago's aquifers.

Sixteen megawatersheds were delineated by digital composite analysis of the geological map, inferred fracture zone maps, and hydrological data on a DEM to classify those areas most likely to possess a combination of extensive open fracture zones, favorable lithologic boundaries, and optimum recharge from high mountain rainfall or

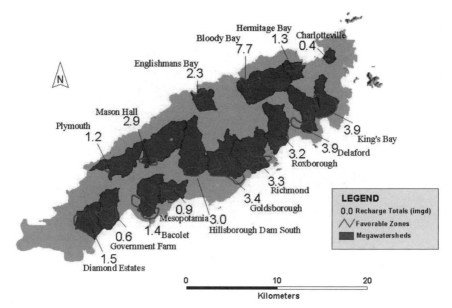

Figure 42. Favorable zones in Tobago with recharge totals.

TABLE 9. Traditional Aquifers, Megawatersheds, and Tested Well Production at Sites

Aquifer	Megawatershed Model Recharge, mgd	Alluvial Model Recharge, mgd	Tested Production Capacity, mgd
Diamond Estates	1.5	0.0	1.52
Government Farms	0.6	0.24	0.63
Bacolet	1.4	0.19	0.88
Belmont	3.0	0.0	0.68
Bloody Bay	7.7	0.33	0.36
Charlotteville	0.4	0.05	0.16

runoff along stream valleys. Fifteen sites within megawatersheds were identified as "favorable zones" possessing the greatest number of these favorable attributes and in closest proximity to WASA pipelines. Figure 43 illustrates a matrix approach to screening for favorable zones within megawatersheds.

Detailed geological and geophysical surveys were then carried out in the favorable zones. Survey data were integrated into the mega-watersheds models, and test-drilling sites were selected based on the highest potential benefit/cost to WASA.

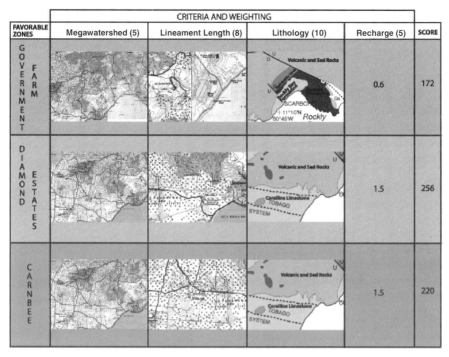

FAVORABLE ZONES	CRITERIA AND WEIGHTING				SCORE
	Megawatershed (5)	Lineament Length (8)	Lithology (10)	Recharge (5)	
GOVERNMENT FARM				0.6	172
DIAMOND ESTATES				1.5	256
CARNBEE				1.5	220

Figure 43. ETI/HSA-Lennox Tobago Groundwater Exploration Program, favorable zones criteria matrix example. (See color insert.)

MODEL CALIBRATION

Solving the island's water supply problems required an additional 3–5 million gallons per day of reliable water-into-supply. This would not only reduce the overall deficit, but also would increase the reliability of supply. Without the megawatershed paradigm, there would not be an expectation of producing significant additional groundwater on the island. Recharge figures shown in Table 9 imply that if this new model is applicable, then substantial new water is available. In actuality, during the project executed on the island, the contractor produced 4 million gallons of water per day at strategic locations along the island's water distribution network. Wells drilled within several of the defined megawatersheds produced over a long period of time quantities of water consistent with the predictions of this model, and additional production wells could be constructed in several of the calibrated megawatersheds such as Bacolet, Belmont, and Bloody Bay. Testing of the wells predicted the capacity of the wells and aquifers [Ingari et al., 2000]. Sample constant rate test results are shown in Figure 44, and

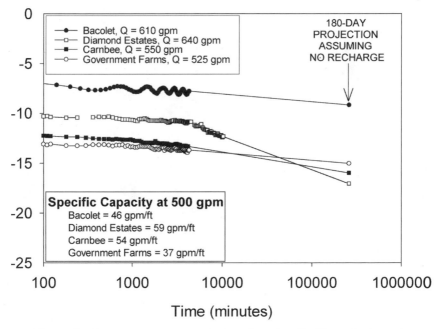

Figure 44. Constant rate test results for selected wells. (See color insert.)

TABLE 10. Field Observations at Current Wells

Aquifer	Well Name & No.	Production mgd	June (2001) Water Levels Meters, (MSL)	Critical Water Levels Meters, (MSL)
Diamond Estates	Diamond Estates #1	0.3	4.38	−16.35
	Diamond Estates #2	0.4	3.52	−0.22
	Carnbee #1	0.57	2.54	−35.24
Government Farms	Government Farms #5	0.55	3.11	−63.38
Charlotteville	Charlotteville #2	0.08	−2.80	−5.64
Mason Hall	Belmont Road #1	0.57	19.6	10.70

recommended production levels of the individual aquifers are shown in Table 10.

Several of the production wells have been pumping into the distribution system for over a year. During that period of time, over

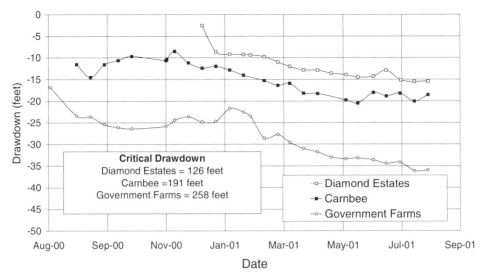

Figure 45. Water levels in wells after one year of pumping.

575 million gallons have been pumped from the Diamond estates and Government Farms megawatersheds. Figure 45 shows the water levels in the wells over the dry season and the start of the wet season. From this figure, it is apparent that the water levels in all the wells had a steady decline during the dry season and are either leveling out or increasing during the start of the wet season, indicating that active recharge is occurring and validating the megawatershed model. If megawatersheds did not exist, then the water levels would have rapidly declined and would have not recovered during the wet season.

Validation of the estimate of groundwater recharge will take several years of long-term monitoring. However, preliminary results based on water-level trends and pumping rates seem to indicate the quantity of recharge estimated is within the margin of error.

Evapotranspiration

The Crown Point airport in the low-lying area of the southwest of the island provides estimates of free surface evaporation that was calculated by using the Penman equation. As proposed by Penman for tropical regions, these data were multiplied by a factor of 0.75 to estimate actual evapotranspiration on a monthly basis. The evapotranspiration data from the Crown Point airport do not represent the evapotranspiration for the whole island, especially because the Crown Point area receives the least rainfall of the whole island. Consequently, the interior of the island has a greater period of cloud cover and is cooler due to a higher elevation. As solar radiation and temperature are important variables that impact the amount of evapotranspiration, areas with greater rainfall should have a lower evapotranspiration. The inverse relationship obtained from a linear regression of the data yields an equation:

$$\text{Evapotranspiration} = -0.146 * \text{Precipitation} + 1276$$

that was used to compute annual evapotranspiration from rainfall data.

Most basins in Tobago are underlain by similar soils and have similar topography; therefore, a graph of annual runoff versus annual precipitation for several basins should be statistically significant and should allow a relationship between precipitation and runoff to be developed that could be applied over the entire island with a high degree of confidence as shown in the plot. A least-squares linear regression of this data results in the following equation relating annual runoff to annual precipitation:

$$\text{Runoff} = 0.4 * (\text{Precipitation}) - 334$$

that was used to calculate annual runoff from precipitation.

These relationships were applied to the 30-year isohyetal map available for the island such that a recharge from the water balance equation computed and a contour map of the entire island was developed. The recharge map is shown in Figure 41. Also shown in the figure is the topographic expression of the island into minute sub-basins, GRUs, into which recharge can be computed either within the traditional or the megawatershed aquifers by overlay and integration with ArcView. The approximately 1-km^2 GRUs were delineated using a proprietary software package. The resultant recharge map was digitized into ArcView to be used for calculating the recharge to each GRU and therefore the recharge into the entire megawatersheds.

FAVORABLE ZONE ASSESSMENT

Favorable zones for potential wellfield development were determined through an assessment of the various areas with respect to inferred fracture zone direction and length, lithology, and recharge to determine their potential as aquifers. The megawatersheds that feed the down gradient favorable zones delineated by the ETI teams' structural analysis of the fracture architecture of the island were used as the recharge areas (Figure 42).

The lithology of water-bearing rock affects its response to stress (e.g., brittle vs. ductile deformation), primary and fracture porosity, and chemistry of the groundwater. Lithologies high in calcium and magnesium carbonate or rock with veins of the carbonates intermingled will impact on water hardness. Rock with minerals such as pyrite and glucophane can have high levels of iron in the groundwater. The geological mapping of the island and chemical analyses of the few wells and springs on the island suggested an absence of carbonates and iron-producing minerals. The major impact of the lithology on groundwater potential is associated with the permeability of the associated rock and its ductility (the more brittle rocks will likely fracture more easily and have greater permeability).

The formations considered to have the highest rankings include the very porous and permeable Booby Point coral deposits, and the Rockly Bay formation because both units are producers of groundwater either through wells or springs. The Parlatuvier and Mt. Dillon formations are ranked high because they are more silica-rich than other formations and are older; thus, they have been fractured by more periods of tectonic activity.

The medium-ranked formations can all be grouped as the youngest volcanic sequence with the exception of the biotite tonalite, which is the most silica-rich igneous rock on the island. These volcanic rocks have some interbedded sedimentary rocks, which, if intersected at depth, could be prolific groundwater producers. However, no wells were drilled into these formations and it was not known whether prolific strata bound aquifers exist; therefore, these formations were ranked as medium.

The lower-ranked formations generally consist of more mafic and ductilely deformed lithologies. As mentioned earlier, the more mafic a lithology, the higher the proportion of clay minerals produced from each volume of bedrock destroyed, This higher clay content can plug fractures, resulting in a substantially lower permeability. In addition, when a formation is significantly deformed during ductile deformation, it is less likely that subsequent brittle deformation will result in continuous open fracture zones. The contorted, preexisting ductile fabric will absorb the tectonic movement without creating new fractures. Consequently, the exploration team considered the mafic hornblende schist, ultra-mafic rocks, mafic gabbro diorite, and the extensively deformed gabbro-diorite rocks to be less favorable and therefore received a lower ranking.

The location and distribution of geologic structures is also critical in regard to favorability of an area for groundwater development.

In crystalline rock terrains, bedrock aquifers derive most of their porosity and permeability from faulting and fracturing in the bedrock. Fault and fracture zones can also serve to increase the storage capacity and transmissivity of a formation. The regional tectonic regime usually dictates the nature of fault and fracture zones. Tensional (open-standing) zones are the most conducive to groundwater flow. ETIs structural mapping and tectonic studies reveal faults and fractures in Tobago that trend from 72° to 105°, and they are tensional and open-standing. Therefore, many such zones are capable of storing and transporting large quantities of water. Wells that tap these structures can have high yields in otherwise low-yield rock formations. In addition, it was determined that mapped northwest-trending fault zones are also tensional in nature. Thus, fractures or faults oriented in that direction (between 315° and 345°) are also likely to be open and water bearing. Structure length and fault azimuth were used as structural parameters for assessment of aquifers.

Groundwater availability, or recharge, is an important criterion for estimating groundwater development potential. The greater the recharge to a favorable zone, the higher ranked the aquifer. This would result from higher infiltration ratios or larger areal coverage of megawatersheds supplying groundwater to each zone. In Tobago, these categories ranged from less than 1 imgd up to more than 9 imgd.

Four parameters, lithology, fracture length, fracture azimuth, and recharge, were assigned predetermined values based on aquifer favorability, with which the 16 favorable zones shown in Figure 38 were assessed. Figure 42 shows the recharge computed and the ranking of each of the favorable zones conceptualized on the island.

The favorable zone map and the distribution system layer were combined by GIS to determine which of the most favorable bedrock and alluvial aquifers are located within 1 km distance of the existing water authority's transmission mains. The potential bedrock aquifers with the highest ranking were selected for detailed exploration activities, including the estimations of groundwater recharge. Quantitative analyses of some of the highest-ranked parts of megawatersheds were constrained by WASA's requirement that production well targets be proximal to the very limited existing infrastructure and to local consumer centers.

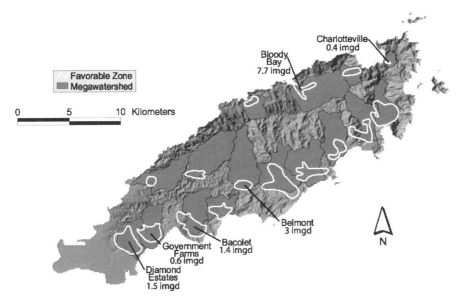

Figure 46. Hydrogeological (megawatersheds) map of Tobago.

Conclusions—Verification of Concepts

The reinterpreted hydrogeological (megawatersheds) map of Tobago (Figure 46) indicates a newly identified class of groundwater environment for Tobago, megawatersheds, identified using a unique exploration program. This map also represents a substantial upgrade of structural geology of a 1998 Geological Map of Tobago. Sixteen megawatersheds were identified, each with an associated "favorable zone," targeted for least-client cost, highest client benefit water well production. Innovative hydrological analysis and modeling was employed using the limited rainfall and stream flow data to maximum value.[4] These were used to calculate runoff, evapotranspiration, and groundwater recharge into megawatersheds totaling 66 MCM/Year (40 MGD) of renewable fresh water resource in the deep bedrock of the island, a 50-fold increase over previous estimates of groundwater availability using traditional hydrogeological concepts and development technologies.

Of the 16 favorable zones identified, 6 of these were drilled and tested on the island and are producing close to 4 MGD of potable water. Nearly three years of production well pumping and monitoring through the two worst droughts on record have effectively verified and calibrated the megawatersheds models of these areas and point toward the voracity of the remaining megawatersheds models on Tobago. In fact, as several wells on the island were drilled and tested for their water potential, an assessment of the model can be made by comparison of the computed recharge and water production potential of the various sites by application of the traditional alluvial model to that using the megawatershed model and by further comparison with the actual tested production of the aquifers.

These comparisons are shown in Table 9 demonstrating the dramatic underestimation of groundwater potential and production by pre-viously applied, generally accepted conventional hydrogeological concepts to the nature, flow, and recharge of aquifers on Tobago. It should be noted that at Diamond Estates, Government Farm, and Charlotteville, the demand for water allowed for the drilling of wells that would maximize their water production, thus allowing a good test for the reliability of the megawatershed model. The other three areas demand requirements that were met with only portions of the full megawatershed potential developed. Notwithstanding, even these wells all produce substantially more sustainable water than predicted by the traditional alluvial model.

Pumping tests were conducted on the wells drilled and plotted according the modified Jacob equation that describes drawdown levels under nonequilibrium conditions at constant discharge rate within a radial aquifer. The following equation:

$$\text{Drawdown} = 264 * \text{Discharge Rate}/\text{Transmissivity}$$

is expected to yield a straight line, of which the slope can be used to determine an "apparent transmissivity." Values obtained for the wells drilled are as high as $160,000 \text{igpd/foot}^2$. These wells behave as large radial aquifers for which boundary conditions were not attained for the duration of the tests. Representative constant-rate (three days) testing results of several of the wells drilled into selected aquifers are shown in Figure 47. Even without further analysis, it is apparent that these bedrock wells have extremely high specific capacities, ranging from 37 to 59 gallons per minute per foot-drawdown.

The extrapolations to 180 days to simulate an entire dry season with no recharge indicate that the drawdown on these wells is very low (less than 2 m), indicating very sustainable aquifers within extensive mega-watershed catchments.

Two of the aquifers have been in production at their tested poten-tial for over three years. These are Diamond Estates and Government Farms favorable zones for which recharge had been predicted to be

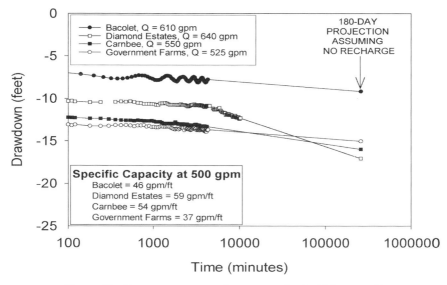

Figure 47. Constant-rate pumping test results for Tobago wells.

2.4 MCM/Year (1.5 mgd) and 0.9 MCM/Year (0.6 mgd), respectively. Both of these aquifers produced on a sustained basis at 80% of their calculated recharge limits. Monitoring of the water levels in the four wells has indicated drawdown levels well within their predicted critical drawdown limits. During the years 2001 and 2002, the island experienced its driest dry seasons in the 50 years of hydrological record. Year 2001 Monitoring data on these wells shown in Figure 48 has continued through 2002 and demonstrates that they have stood up to the driest seasons on record and have shown signs of water well recovery with the onset of the rainy seasons each year. The maximum drawdown at Government Farms was substantially less than the critical drawdown. Monitoring of production and observation wells for water quality to assess the possibility of salt-water intrusion, has, during a record drought year, indicated no evidence of potential for contamination in these coastal bedrock aquifers. Summary data on the year 2001 data on the status of the wells are shown in Table 11.

The groundwater from all wells drilled meets and exceeds all WHO standards for potable water quality, even the aesthetic standards. Water quality data upon handover of the wells and during the production history of each of the wells have confirmed that the source of groundwater is from very stable rock environments demonstrating insignificant changes in water characteristics over time.

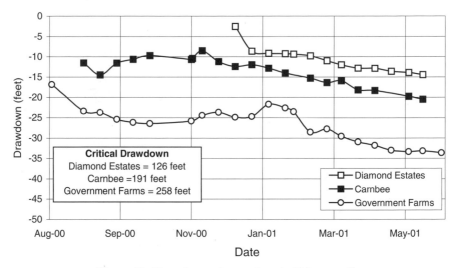

Figure 48. Drawdown observations in Tobago wells.

TABLE 11. Summary Data on the Year 2001 Data on the Status of 4 Tobago Bedrock Production Wells

Aquifer	Yield imgd	Water level, m MSL	Critical Water Level, m MSL
Diamond E	0.3	4.38	−16.35
(3 Wells)	0.4	3.52	−0.22
	0.57	2.54	−35.24
Govt. Farm	0.55	3.11	−63.38
Charlotteville	0.08	−2.80	−5.64
Mason Hall	0.57	19.6	10.7

CONCLUSIONS

The World Bank and United Nations report that over much of the globe, human demand will exceed known sources of natural fresh water supply early in the next century.[5] It was estimated that currently one-tenth of the world's population does not have access to adequate potable water, and by the year 2025, this proportion will increase to one-third.[6] This is equivalent to 2.8 billion people. Solutions to the water supply-demand situation require new ways of doing business using new technologies to solve long-standing problems. The Tobago Megawatersheds Exploration and Development Project was designed as a private-public partnership using highly innovative technologies.

The results of the Project transformed Tobago from a desperately water-constrained to an water-driven economy within a 12-month period, dramatically changing the perspective of government planners and private developers, who now see economic prosperity as a realistic, near-term goal for this small island developing state.

BIOGRAPHY

Dr. Utam S. Maharaj trained as an organic chemist and petroleum engineer and has pursued a career in technological research and the applied sciences. As Head of Research at Petrotrin, Trinidad's national petroleum company, he was at the forefront of seeking and applying novel technologies to solve oilfield problems and the conversion of uneconomic resources to petroleum reserves. In the 1990s, the Water and Sewerage Authority of Trinidad and Tobago (WASA) requested that Petrotrin allow Dr. Maharaj to accept a temporary management position at WASA to help build a management and technical team capable of utilizing state-of-the-art technologies to solve long-standing water-supply shortages in the twin-island Republic. He accepted the challenging role of director of WASA's Water Resources Agency and, during his three year tenure, successfully built a highly motivated water resources team, brought new technologies into WASA, and implemented a historical pilot groundwater development project that solved Tobago's 40-year water crisis within twelve months and provided a model for later success in Trinidad. In 2002, Dr. Maharaj returned to Petrotrin, where he is engaged in assessing microwave-enhanced technology to upgrade heavy petroleum. He has published extensively in both the petroleum and water literature.

REFERENCES

1. http://www.flora.org/flora.mai-not/3528, Austext International transcript of UN Paris conference on managing the world's limited fresh water supplies, March 23, 1998.
2. "Solutions for a Water Short World" by D. Hinrichsen, B. Robey and U. Upadhyay, in Population Reports, Series M #14, Baltimore, John Hopkins University School of Public Health, 1998.
3. "Water Resources Management Strategy for Trinidad and Tobago," by DHV Consultants, Delft Hydraulics and Lee Young and Partners for the Ministry of Planning and Development, Trinidad and Tobago, 2000.

4. "The Megawatershed Concept and Recharge Analyses-Sustainable Groundwater Extraction from Bedrock Aquifers in Tobago," by U.S. Maharaj, R. Sankar, R. Hoag, Jr., R. Bisson, and J. Ingari, presented at the Caribbean Water and Waste International Conference, Port-of-Spain, Trinidad, 2000.

5. http://www.flora.org/flora.mai-not/3528, Austext International transcript of UN Paris conference on managing the world's limited fresh water supplies, March 23, 1998.

6. "Solutions for a Water Short World," by D. Hinrichsen, B. Robey, and U. Upadhyay, in Population Reports, Series M #14, Baltimore, John Hopkins University School of Public Health, 1998.

6 Case Study—Trinidad, West Indies 2000–2002

INTRODUCTION

This chapter is based on drawings, specifications, and other documents made by and on behalf of Earthwater Technology Trinidad and Tobago, LLC, during an extensive groundwater exploration and development project carried out from 2000 to 2002. The project represented a true public-private partnership and brought the best available foreign technologies to Trinidad, uniting U.S. and a local Caribbean business to achieve unprecedented groundwater discoveries exceeding 200 imgd of untapped potential previously believed to be impossible. Trinidad transitioned from a seemingly seawater-desalination-dependent island to true water independence in 24 months.

The project was based on the year 2000 success in Tobago of the novel megawatersheds exploration approach to groundwater discovery.

The contractor who performed the work for the Water and Sewerage Authority of Trinidad & Tobago (WASA) was a U.S.-Trinidad Joint Venture of ETI and Lennox Petroleum Services Ltd. (Lennox). ETI brought new scientific concepts and technologies to Trinidad, performed the Hydrogeological Reassessment, and provided overall project supervision, well siting, design and drilling direction, plus production well and aquifer testing. Lennox provided state-of-the-art drilling equipment, all well drilling services, and capital works construction. Although groundwater exploration covered the entire island and identified several hundred million gallons per day of untapped sources, the contract limited the actual development of new water sources to within a few kilometers of existing WASA pipelines (cumulative distance not to exceed 14 km), which in hindsight was ill-

Modern Groundwater Exploration: Discovering New Water Resources in Consolidated Rocks Using Innovative Hydrogeologic Concepts, Exploration, Drilling, Aquifer Testing, and Management Methods, by Robert A. Bisson and Jay H. Lehr
ISBN 0-471-06460-2   Copyright © 2004 John Wiley & Sons, Inc.

considered because numerous major groundwater sources were identified outside the distance limits and old pipelines could not handle water quantities from new water sources averaging 1 imgd each. Although this restriction caused numerous delays and extra costs to the contractor, the ETI-Lennox JV actually delivered 15 imgd within a collective distance of 11 km of existing mains.

The results of the 2002 Trinidad Hydrogeological Reassessment represented a true paradigm shift in the country's vision of its fresh water resources, from continued chronic dependence on unreliable, capital and time-consuming surface water sources to dependable, inexpensive, fast-track groundwater sources. ETI identified sustainable groundwater resources in previously unobserved megawatersheds, aquifer systems, and aquifers in consolidated rock and unconsolidated sediments exceeding 280 million imperial gallons per day (imgd), whereas previous estimates limited combined new production potential on Trinidad and Tobago to 17 imgd.

Newly discovered bedrock megawatersheds, aquifer systems, and alluvial aquifers in Trinidad were verified and calibrated during production well construction, which yielded more than 15 million imgd of new, potable groundwater supply in accord with the Contractor's Agreement with WASA. Within a three-year time frame, the amount of new fresh water developed from megawatersheds projects carried out for WASA on the islands of Trinidad and Tobago exceeded 20 million imperial gallons per day (imgd) at a total capital cost to WASA of less than $30 million and an operating and maintenance (O&M) cost of less than $0.40 per 1000 gallons.

The benefits of the Water Resource Agency's innovative water development programs were evident during the year 2000 and 2003 droughts in Tobago, when surface supplies were severely curtailed and the megawatersheds wells became the primary source of water for the island's indigenous and tourist populations. In Trinidad, the 2003 record dry season threatened surface reservoirs, and the new deep wells helped save consumers from water rationing.

TECHNICAL PROGRAM SUMMARY

In the quest to accomplish a megawatersheds exploration program representing a comprehensive and accurate groundwater assessment, ETI initiated data searches on the broadest possible front and applied the most advanced digital database construction, management, and computer-visualization technologies available. The team collected and digitized diverse data never before pooled, performing analyses of regional tectonic models and satellite remote sensing, digitized over 15,000 data points from oil and gas seismic lines, carried out hundreds of line-kilometers of geophysical surveys, interpreted 1600 hundred logs from oil, gas, and water wells, and completed intensive on-ground geological mapping in prospective megawatersheds. The team then employed its novel "multiple convergent datasets" method of data processing and integration to produce a revised hydrogeological map of Trinidad (Figure 49), significantly upgrading the old groundwater map of the island; identifying several major new sedimentary and bedrock aquifers, aquifer systems, and megawatersheds; and increasing the amount of Trinidad's safely extractable groundwater more than 20-fold.

The unprecedented amount of new digital hydrogeological and groundwater-related geophysical data acquired during the project was integrated with massive amounts of preexisting data from WASA's files and the oil and gas industry, all newly digitized by the team, and then combined with revised interpretations of geotectonic framework and geology to create a large-scale new geological map of Trinidad (Figure 50). ETI published the new map and associated cross sections using state-of-the-art data integration and visualization software to merge new discoveries with other investigators' recent published and un-published works. ETI provided printed and digital copies of the new Geological Map to WASA for dissemination to other government ministries. The ETI 2002 Geological Map represents the most compre-hensive geological update available of the structural maps and basin stratigraphy of Trinidad's northern and central range and adjacent basins.

The results of the 2002 Trinidad megawatersheds project were reported in a four-volume printed report, which contains hundreds of original, high-resolution, full-color maps and cross sections of Trinidad's geology and hydrogeological environments. Trinidad's Water Authority was also provided with a digital, CD version of the report and illustrations as well as new, large-scale geological and hydrogeological maps of Trinidad, high-resolution satellite image-maps, and over CDs with over three gigabytes of GIS-referenced, digital data supporting the findings reported by the ETI technical team.

Figure 49. Hydrogeological map of Trinidad—2002 (Earthwater Technology Trinidad & Tobago LLC). (See color insert.)

194

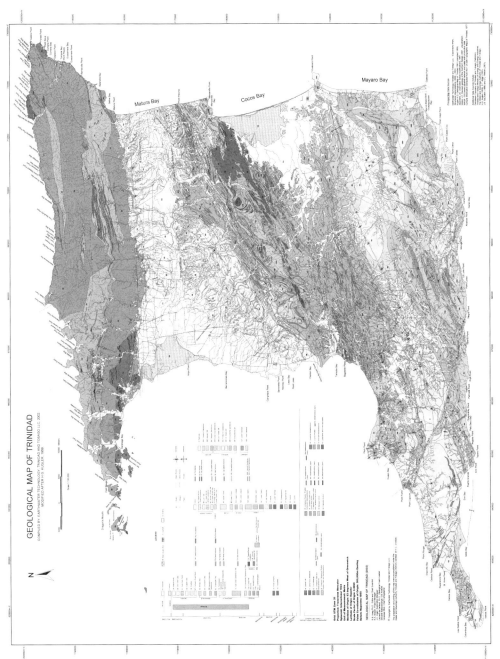

Figure 50. Geological map of Trinidad—2002 (Earthwater Technology Trinidad & Tobago LLC). (See color insert.)

195

PROJECT TEAM

The project team comprised a diverse and committed group of U.S. and Trinidadian private and public sector managers, scientists, technologists, and administrators, who collectively were ultimately responsible for achieving the unprecedented goals of the project.

The Earthwater Technology—Lennox Petroleum Trinidad Hydrogeological Reassessment and Well Development Project Joint Venture team was led by ETI president and project director Robert A. Bisson, principal scientist Dr. Roland B. Hoag, field exploration manager and chief geophysicist Joseph C. Ingari, chief geologist Dr. Laurent de Verteuil, LPSL CEO Wayne Parsad, drilling program manager Lennox A. Persad, chief driller Carlisle Charles, and drilling logistics manager Sieu Rambhajan. Michael Sununu of JHS Associates administered ETI's financial operations.

An eclectic group of professional scientists and technologists worked both in Trinidad and the United States from firms including Earthwater Technology International, Inc., HydroSource Associates Inc., JHS Associates Ltd., Lennox Petroleum Services Ltd., U.S. Filter Operating Services Company, Earth Satellite Corporation, Earth Science Applications Systems, and other subcontractors. Key team members included assistant field exploration manager David Hisz and geoscientists John Weber, Cameron Warlick, Chad Wolak, Andy McCarthy, Fred Bickford, Wayne Randall, Rob Otto, Jeremy Coerper, Christine Bowman, Walter Ebaugh, Mark Jancin, David Yoxtheimer, as well as chief cartographer Shivanti Balkaransingh, GIS specialist Venessa Elliot, and in-country logistics administrator of Lennox Offshore. Venture partner in-country office managers Carol Balgobin and Virla Bally and chef Carol Darmanie completed the basic team construct.

The genuine commitment and round-the-clock support of WASA's land-man, Leon Toppin, and hydrogeological technician, Randolph Sankar, were particularly critical to the ultimate success of the program. During the project, many of WASA's scientists, engineers, and technicians became part of the team as they were trained in megawatersheds investigative, drilling, testing, and management technologies by ETI and LPSL specialists in U.S.- and Trinidad-based classrooms and in the field. The challenges of shifting political environments, sweeping changes in WASA management and unexpected delays were ably tackled by JHS Associates president Dr. John H. Sununu, whose thorough understanding and ability to articulate everything from megawatersheds technological esoterica to the colossal socioeconomic and political benefits derivable from a successful project, kept key governmental and financial institutional players engaged as partners to the final, triumphant conclusion of the venture.

The Hydrogeological Reassessment Report and accompanying maps and digital databases were produced for Earthwater Technology Trinidad & Tobago LLC by its principal subcontractors Earthwater Technology International, Inc., HydroSource Associates, Inc., Latinum Ltd., and Earth Satellite Corporation, with administrative oversight by JHS Associates Ltd. and data access and personnel support from the Petroleum Company of Trinidad and Tobago Ltd.; Lands and Surveys Division of the Ministry of Planning; Department of Surveying and Land Information, University of the West Indies; Ministry of Energy and Energy Industries and National Quarries.

The successful implementation of this novel and challenging "results-based" contract also involved close teamwork and financial risks and strong financial support by the prime contractors, subcontractors, ETI's U.S. partner and financier U.S. Filter/Vivendi USA, and LPSL's lender Republic Bank of Trinidad.

Setting

Trinidad, West Indies, lies in the extreme southeastern Caribbean Sea, 11 Km northeast of Venezuela. Trinidad is bordered by the Caribbean Sea to the north, Atlantic Ocean to the east, Columbus Channel and Venezuela to the south, and Gulf of Paria and Venezuela to the west (Figure 51). The Island is approximately $4800 \, km^2$ and is characterized primarily by plains and low mountains. Physiographic divisions include the Northern Range, Northern Basin, Central Range, Southern Basin, and Southern Range.

The Northern Range of Trinidad is the easternmost expression of the Andean mountain chain and consists of a series of metamorphic formations that include Trinidad's highest mountain El Cerro del Aripo, 3085 feet. This mountainous range is covered with various types of forests, starting with Littoral Woodlands along the coastlines, Evergreen Seasonal Tropical Forests in the lower elevations, Lower Montane Rainforest in elevations from 500 to 2500 feet, Upper Montane Rainforest in elevations above 2500 feet, and Elfin Woodlands covering the highest mountains' peaks. The Northern slopes meet the Caribbean Sea, and the southern slopes end in the Caroni plains. Trinidad also possesses two significant lowland tropical wetland

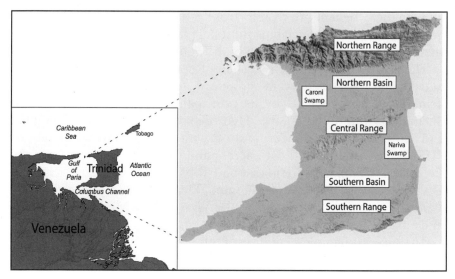

Figure 51. General location and physiographic map of Trinidad. (See color insert.)

systems, the Caroni and Nariva swamps, bordering the western and eastern coasts, respectively.

Trinidad's climate is tropical and characterized by two major seasons: a dry season from January to May and a wet season from June to December. A short dry spell, "Petite Careme," typically occurs in the middle of the wet season in September or October [Granger, 1982; Water Resources Agency (WRA), 1990]. The average annual temperature is approximately 26°C with minor diurnal variations [WRA, 1990]. Granger [1982] and the WRA of Trinidad and Tobago [1990] report an average annual rainfall of approximately 2000 mm with over 78% of the mean annual rainfall occurring during the wet season.

Because of the prevailing northeast trade winds and orographic effects, the highland areas of northeast Trinidad receive up to 3800 mm per year of rainfall [Granger, 1982; WRA, 1990]. Generally, the amount of precipitation ranges from 1250 mm in the northwest and southwest of the island to greater than 3000 mm in the northeast [Garstang, 1959]. Records indicate up to 60% of the total rainfall received in some areas is lost to evapotranspiration [WRA, 1990].

Background of Groundwater Studies in Trinidad

Numerous individuals, government agencies, and private consultants have studied the geology and hydrology of Trinidad since the mid nineteenth century. A description of the most important of these investigations, reports, and maps is included further on in this case study.

In an attempt to collect, evaluate, and summarize the plethora of previous works with relation to groundwater potential in the twin-island republic, WASA in 1997 contracted with DHV Consultants BV in association with Delft Hydraulics and Lee Young & Partners (DHV) to carry out a two-year program of compilation and evaluation of all available previous Trinidad and Tobago water studies. In June 1999, a summary of results was presented in a report entitled "Water Resources Management Strategy for Trinidad and Tobago." The DHV Report included specific recommendations to alleviate future water shortages by taking both demand- and supply-side measures.

On the groundwater supply side, DHV's evaluation team found they could not use groundwater modeling as a tool for estimating safe yields because of inadequate aquifer data, and it recommended that future, proper hydrogeological investigations should be carried out to generate reliable figures for recharge and safe yield. From all prior reports, DHV estimated combined groundwater potential in Trinidad and Tobago at approximately 69 million imperial gallons per day (imgd), of which only 17 imgd remained available for additional extraction from aquifers, mostly in the Southern Sands of Trinidad and none in bedrock. Groundwater was therefore considered a local resource that would play no further role in Tobago and only a minor role in any Trinidad future water supply strategy. Surface waters were considered the only viable supply-side alternatives in Trinidad, including construction of a new dam in Tobago's rainforest, and in Trinidad a seawater desalination plant at Point Lisas and dams in North Oropouche and Inniss.

CONSTRAINTS AND CHALLENGES

Introduction

Although enjoying considerable "good will" from optimistic WASA management and technical personnel, the Trinidad megawatershed exploration and development project was implemented during a time of great political, economic, and climatological upheaval, making the path to success very rocky indeed. During the term of this project, Trinidad underwent huge weather extremes and two changes in government; WASA was completely reorganized; and the United Stated was attacked by terrorists on an unprecedented scale. These unforeseen events had profound, negatively synergistic impacts on the new ground-water project.

Dangerous Wildlife

Unlike Tobago's benign rainforest, Trinidad's wilderness hosted myriad toxic fauna and flora, and ETI's chief scientist, Dr. Roland "Skip" Hoag, had a biting encounter with the top of the local reptilian food chain, a fir de lance. In a remarkably lucky amalgamation of circumstances, Hoag was in the field with his long-time colleague Joseph Ingari, who quickly and correctly administered first aid and used his cell phone to alert Bisson in Port of Spain. Within minutes, drilling partner Lennox Persad had a helicopter en route to the remote location, and Hoag was soon deposited in the midst of an astonished group of soccer players on the Queen's Savannah across from St Claire Hospital in Port of Spain. Hoag was provided excellent emergency treatment by Dr. Earl Brewster, a prominent gynecologist, and then flown to Miami via a medical jet for specialized treatment; he eventually made a full recovery. However, this single event caused substantial lost time for key explorationists at a crucial time in the project.

Data Processing and Communications Links

The amount of data being collected, analyzed, and integrated by the team required 15 networked computers at the ETI Trinidad offices. In order to accomplish these tasks, close coordination and data sharing with U.S.-based teammates was essential, which required massive digital data transfer capabilities. A combination of telephone line limitations and outages and electricity voltage swings, outages, and power surges made the practice of such communications unreliable, so time-consuming CDs by courier were necessarily extensively used as supplemental tools.

Team Solutions

Reflecting the characteristic good will and camaraderie developed between client and contractor during the different, but equally taxing, Tobago project, WASA and ETI-LPSL management worked fever-ously to adjust to these conflicting forces through weekly and semi-weekly meetings, urgent e-mailing, telephoning, and faxing. Myriad compromises were made to permit continued progress under seemingly impracticable conditions.

The ETI-LPSL JV and all of the subcontractors went far beyond the call of duty in a fierce show of commitment to delivering water to desperate consumers as soon as possible. The result was a win-win for the Government, the Water Authority, the Contractor, and most of all, the people of Trinidad, who gained 15 million new gallons per day and the assurance of hundreds of millions more as-yet untapped indigenous, renewable treasure that has the potential to make energy-independent Trinidad self-sufficient in water and food as well.

TRINIDAD TECHNICAL PROGRAM: CONCEPTS, METHODS, AND TECHNOLOGIES

Introduction and Overview

Groundwater environments in Trinidad are locally extremely complex and varied throughout the island. Trinidad's physiographic setting is dramatically different from neighboring Tobago's, and geologists, hydrologists, and hydrogeologists using traditional models and methods had studied Trinidad for much of the nineteenth and twentieth centuries with very limited results as reported in DHV Consultants' 1999 summary report containing negative conclusions about potential additional undeveloped groundwater sources, especially in the Northern Range. The ETI 2000–2002 hydrogeological reassessment thoroughly tested the efficacy of employing the novel megawatershed paradigm and exploration approach against the backdrop of such conventional models.

The Trinidad megawatershed exploration and groundwater development program incorporated twenty-first century geological theory with space-age geodesy, economic minerals exploration, and water well drilling technologies.

The team implemented the exploration program using remote sensing, surface, and subsurface field data collection, and geographic information systems (GIS) data integration and analysis by a highly experienced interdisciplinary team who systematically applied multiple techniques and technologies with a long history of development by the team leadership. The Trinidad megawatershed exploration program further advanced the state-of-the-art by incorporating a rigorous recharge model based on multisource hydrometeorological data, including satellite sensors and field-verified hydrogeological attributes. The output from this model was input into a GIS database containing detailed surface geological and geophysical data to quantify ground-water potential on the island.

A brief explanation on the methods, techniques, and technologies used during this study follows. Intended for a broad readership, "terms of the art" or jargon are used only where essential, but this is unfortunately frequent because the terms define specific instruments, models, software, or other tools of a highly specialized exploration approach. Internet search engines can provide quick explanations of the more arcane terms.

Also, for the advanced researchers reading this book, an extensive listing of metadata was included in Vol. IV, Appendix XIII of the full technical report available at WASA and the University of the West Indies Library, documenting USGS protocols, all details of data sources, vintages, and related information that serve to validate the data.

The full report contains a detailed section on the estimation of aquifer properties, primarily hydraulic conductivity (K) from existing well data, and describes the limitations of WASA's pre-ETI data set, for pumping tests in particular. For example, test and constant rate pumping test data provide the best means for evaluating aquifer performance and average aquifer properties of K and transmissivity (T), but pumping test data were only available for a few wells in Trinidad prior to ETI's work, and the few available step test data far outnumbered more accurate constant rate tests.

The Megawatershed Paradigm includes geology as the major controlling influence in groundwater environments, and so an accurate, digital, GIS-based geological map was a fundamental requirement to launch an effective groundwater exploration. As no digital geological map of Trinidad was available, ETI produced a new digital "working" geological map based on the 1959 1:50,000 Kugler map series. In addition to massive technological issues of a GIS nature, this project also involved many technical editorial decisions for consolidating and revising the lithologic map units among the Kugler 1:50,000, the Kugler 1:100,000, and the Saunders 1:100,000 maps.

During the course of the project, the team substantially modified the new digital basemap to reflect new geological interpretations, especially in the Northern Range and just south of the Northern Range (ETI investigators used seismic data to identify previously undocumented fault systems that extend to surface) as well as in the Northern basin. The Northern Range map represents a major departure from all previous maps, and it was based on the recent and ongoing research and mapping of Dr. John Weber, who worked on this project as part of the ETI team. The new map abandons the earlier approach of mapping stratigraphic formations in the metamorphosed Northern Range rocks in favor of simply delineating (mapping) metamorphic lithologies and structural provinces. This approach was necessary because the original stratigraphy has been obliterated through most of the range, and classic concepts of depositional superposition cannot be applied. The full report contains a description of all existing, new and revised geological units included on the ETI digital geological map.

MEGAWATERSHED PARADIGM

The space age has induced quantum leaps in geological theory, and comparable advances in geophysical instrumentation and interpretive methods have permitted scientists to discover new information about the genesis, composition, and structure of the Earth's crust and the internal dynamics of the many fluids (e.g., magma, oil, water) contained within it. Equally important, the perfecting of computer-based geographic information systems allowed skilled explorationists to quickly and accurately combine and analyze many different types of data, leading to major discoveries of oil, gas, minerals, and, more recently, to breakthroughs in the understanding of the nature and extent of groundwater resources, including the discovery of complex aquifer systems coined "megawatersheds" by the private sector explorers who first documented the phenomenon, circa 1989, in the Great Rift Systems of East Africa [Bisson et al., 1989]. A brief description of the "Megawatershed" Paradigm is included below.

After the discovery of this new class of groundwater domain, proprietary groundwater exploration techniques were refined by these explorers and successfully applied in hundreds of other areas around the globe, including the island of Tobago in 1999–2000, where a 45-year water shortage was resolved in less than one year through the discovery and development of Tobago's previously unmapped and untapped megawatersheds.

The "Megawatershed" Paradigm is a conceptual model describing natural complex systems of water catchments and drainage genetically linked to tectonism and consisting of three-dimensional surface and subsurface "zones" often evidencing active interbasin groundwater transmission in consolidated fractured bedrock and compartmentalization in faulted, deep sediments (Figure 52). This model supersedes

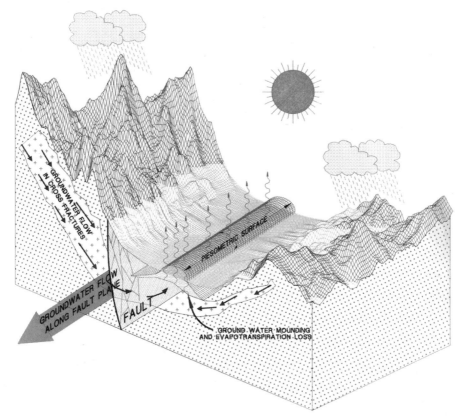

Figure 52. Megawatershed Conceptual Model by Bisson Exploration Services Co. Illustration by Summit Engineering, 1992. (See color insert.)

the traditional watershed model and has profound implications in river basin models, which most often describe topographically controlled, functionally two-dimensional drainages and depict deep groundwater resources as static, poorly recharged artifacts of surface flow, confined to local discreet bedrock or alluvial units.

The Megawatershed Paradigm describes deep groundwater systems' water catchments, recharge, transmission, and storage boundaries that transcend the limits of traditional watersheds and basins by recognizing the overriding influence of tectonically induced, large-scale, highly variable fracture and fault permeabilities and structural geometries in defining rainfall catchments, actual percentages of precipitation contributing to groundwater recharge, and groundwater flow mechanisms in consolidated rocks that frequently transcend synthetic watershed and basin boundaries or, conversely, may possess lateral recharge bound by previously unsuspected stratigraphic offsets also associated with tectonism.

The term megawatershed is used in this case study to refer to sub-surface groundwater systems that are integrated in terms of recharge, discharge, storage, transmissivity, and containment. As mappable groundwater resources, spatially delineated by subsurface structure and stratigraphy, megawatersheds may not coincide with surface topo-graphic divides, and they may receive recharge from parts of several surface watersheds. Similarly, as a structurally contained water resource, wells drilled into multiple aquifers in a single megawatershed may produce water from potentially dissimilar lithologies with common hydraulics related to brittle fracturing and gravity-fed contributions from adjacent and overlying unconsolidated sediments. In fact, mega-watersheds often occupy diverse host environments exhibiting both primary (continuous) and secondary (fracture) porosities with flow systems delineated by fracture systems, faulted or weathered lithologic contacts and igneous intrusions (e.g., dykes), overlying or adjoining weathered rock, and overlying porous unconsolidated sediments. The history and nature of regional tectonic stress fields and principal bounding faults, interacting with lithology and climate, play a major role in determining the geometry and extent of megawatershed boundaries.

The megawatershed concept operates effectively and has scientific validity and economic and management utility at a scale of tens to thou-sands of square kilometers. This study sets a precedent in Trinidad hydrogeology in that it incorporates the serious analysis of subsurface structure and stratigraphy into its evaluation of the island-wide ground-water resource. It is also revolutionary for its first-of-its-kind analysis of precipitation data and resulting recharge evaluation.

MEGAWATERSHEDS EXPLORATION PROGRAM

When the Megawatershed Paradigm is used as a groundwater explo-ration model, as is the case for this Trinidad Project, greatly improved understanding of groundwater environments results, permitting accu-rate, comprehensive assessments of "safe yield" from local, country-wide, or regional water resources. The megawatershed model actually provides a template for measurement of all facets of the hydrologic cycle in reality-based natural hydrogeological catchments, including rainfall, evapotranspiration, and surface and subsurface inflow and outflow interactions with surface water, shallow aquifers, deep fractured bedrock, and deep alluvial aquifers.

The Trinidad Megawatershed Exploration Program further advanced the state of the art by incorporating a rigorous recharge model based on unprecedented hydro-meteorological data quality, density, and duration into a GIS database containing detailed surface and subsurface geological and geophysical data to quantify groundwater development potential on the island.

In a quest to accomplish the most comprehensive and accurate groundwater assessment ever implemented anywhere, the ETI team initiated data searches on the broadest possible front and applied the most advanced digital database construction, management, and computer-visualization technologies available. The team collected and digitized diverse data never before pooled, performing analyses of regional tectonic models and satellite remote sensing, digitized over 15,000 data points from oil and gas seismic lines, carried out hundreds of line-kilometers of geophysical surveys, interpreted 1600 logs from oil, gas, and water wells, and completed intensive on-ground geological mapping in prospective megawatersheds. ETI then employed its novel "multiple convergent datasets" method of data processing and integration to produce a revised geological map of Trinidad that significantly upgrades the current map of the island.

The megawatershed application in Trinidad is a much-expanded version of the ETI Tobago study [2000] and has roots in 1970s studies of regional tectonics and brittle fracturing related to petroleum migration pathways in the U.S. Eastern Overthrust belt using ERTS MSS imagery and LRIS [Bisson and Widger, 1974] and later work using similar techniques for groundwater exploration [Bisson and Widger, 1977; Bisson and Hofman, 1982; Bisson and Long, 1984; Hoag and Ingari, 1985]. Since those pioneering studies were initiated 30 years ago, the concept of GIS-integration of remotely sensed data with other data sources for oil and gas, minerals, and groundwater development has been successfully applied by numerous investigators worldwide.

Database Development, Integration, and Analysis

In order to achieve the goals of the Trinidad Groundwater Reassessment Program, ETI carried out field reconnaissance work and performed a comprehensive review and hydrogeological interpretation of all relevant available material, integrating all into a ETI-digitized version of Kugler's 1:50,000 map sheets, creating a base map upon which the balance of the Project was based. The newly published

Hydrogeological Map of Trinidad includes previously known and just-discovered geological features of the island of Trinidad and locations of well fields and aquifers and socioeconomic information based on the national coordinate systems.

Trinidad's Revised Water Balance

Calculations are included in this Report of water balance and renewable, sustainable groundwater development potential in each megawatershed. These calculations are based on comprehensive hydrological analyses, including ETI's interpretation, using proprietary technologies, of data from multiple satellites, aerial photographic analyses, published geologic, hydrologic, and soil maps, data gathered in the field, and meteorological data from island precipitation stations.

Database Development, The bulk of GIS database development was devoted to data acquisition and quality control. Data types were defined, and a broad search for data sources was launched to discover and access data from every possible source, especially government and oil company archives, where the bulk of previously unanalyzed data resided. Most of the data were in analog format. The most time-consuming step involved acquiring the often proprietary or confidential data from government and private archives and then digitizing and interpreting thousands of analog records for input into the ETI GIS database.

Geodesy. In order to create a database that is useful for spatial analysis using a modern digital GIS, all its parts must be registered to a common coordinate system, but previously published geological and cultural maps of Trinidad were based on a variety of projections and units.

For example, the Cassini Projection Spheroid was adopted for use on maps by Clarke in 1858, and the United States Army Topographic Command subsequently produced a computer-based simultaneous least-squares adjustment of triangulation and traverse called the Naparima Datum of 1972 [Mugnier, 2000], and adopted the Universal Transverse Mercator (UTM) system for use with the new datum. Detailed geological maps from the Natural History Museum Basel [1996] were measured with links and contained no projection, datum, or spheroid information [Cassini, 1858], the 1:50,000 geological map series compiled by Kugler [1959] also uses links and contains no other projection information, and Kugler's 1:100,000 map [1959] uses links with geographic coordinates noted as well (degrees, minutes, seconds). (Note that 8000 links = 1 mile and 1 m = 4.971 links). The 1:25,000 Topographic Map Series [1977] was based on UTM Zone 20, International Spheroid and the Naparima 1955 Datum. However, the petroleum industry retains the Cassini Projection in some applications.

Data conversions between these incompatible sources, particularly when detailed projection information and source documentation were not available, were a potential source of error. Practical use of these geographically inconsistent and sometimes flawed old and recent datasets required thoughtful judgments regarding data precision, and ETI investigators corrected errors deemed critical, while ignoring errors of little importance to Project objectives.

REMOTE SENSING

Aerial Photographs: The 1998 1:25,000 scale aerial photographs of Trinidad were purchased. Photographs of target and field sites were then scanned and georeferenced. The aerial photographs were not orthogonally rectified and could not be registered accurately with Register Image. Thus, the scanned images were georeferenced in Idrisi32 using the Resample module.

The scanned photos were registered in ArcView™ with the Register Image for identification of corresponding points between photo and the Digital Raster Images (DRG). The intersections of road networks were used as control points to create a georeference file for resampling the aerial photos. The east and north of the control points were generated by creating a point shape file of these control points and adding the coordinates with the Add XY ArcView extension. The scanned photos were then imported into Idrisi32, and the corresponding X and Y values were obtained through a manual process. At least four well-distributed points were necessary for resampling. Note that in forested areas, the road network was either sparse or covered by vegetation and it was difficult at times to get properly distributed control points. In most instances, at least seven to ten control points could be identified. The image was then resampled and exported. The metadata of the Idrisi image contained the necessary information to create a world file for each image. Once the image was resampled, a few more control points were identified and the process was repeated.

The greatest discrepancies occurred at the edges and in mountainous regions. Cropping the overlap areas between photographs solved the first problem. The second problem was more challenging, and errors could only be reduced by iteration. The accuracy of the images was on the order of magnitude of ±1.0 m east, ±1.0 m north. Because of the lack of control points on the aerial photographs in the Valencia/Sangre Grande area, the photo mosaic has an accuracy of ±25 m. The accuracies obtained were sufficient for in-field geophysical exploration.

Side Looking Airborne Radar (SLAR) Imagery. SLAR data were acquired from the Ministry of Energy and then scanned, orthorectified, and georeferenced. The image was obtained from Intera Incorporated (now part of Duke Engineering & Services) and used by team company Earth Satellite Cerporation to conduct a structural analysis of lineaments in Trinidad. SLAR was an electronic image-producing system that derives its name from the radar beam transmission being perpendicular to the path of the aircraft during data acquisition. The result was an obliquely illuminated view of the terrain that enhances subtle surface features. SLAR was an active sensor; the system provides its own source of illumination in the form of microwave energy. Consequently, imagery can be obtained either day or night. As microwave energy penetrates most clouds, SLAR can be used to prepare image maps of cloud-covered areas.

LandSat Imagery. Upon completion of the image acquisition and processing, a geological structural analysis was performed, mapping faults, fractures, and folds, using 1:75,000 scale pan-sharpened, multispectral, LandSat 7 Enhanced Thematic Mapper (ETM) satellite image acquired. ETM bands 7, 4, and 2 (R, G, B) were merged with the 15-m LandSat 7 ETM pan band to produce a relatively high-resolution image enhanced for geological structure interpretation. The LandSat 7 ETM data were also processed using the proprietary EarthSat GeoVue™ algorithm to bring out as much lithologic information as possible and was orthorectified and geocoded to the EarthSat GeoCover™ database. This database has a geodetic positional error of less than 50 m.

The interpreted geological features were then digitized into an ARC/INFO® GIS format, written to a CD, and plotted on an electrostatic plotter to provide overlays to the imagery. All of the satellite imagery and the structural analysis were projected into a Transverse Mercator Projection (UTM Zone 14), using the NAD 27 Spheroid and Datum and printed at a scale of 1:75,000. The ETI team used printed and digital versions during the exploration program.

Land cover data were also generated from a LandSat 7 Thematic Mapper satellite image. Unfortunately, clouds obscured some of this image, so the original land cover image was modified using the *Generalize Grid ArcView extension Remove Noise* function, digitally replacing obscured cells with "nearest-neighbor" cells. Land cover was categorized into nine classes: forest/evergreen, shrub/scrub, grasslands, barren/sparsely vegetated, urban, agricultural, permanent/herbaceous wetland, mangrove wetland, and water.

RadarSat Imagery. The August 14, 1977 RadarSat image was geocoded and orthorectified by RadarSat International using 30-m DEM and 1:25,000-scale DRG of Trinidad. Subsequently, 1:100,000- and 1:75,000-scale RadarSat image maps created from this data were produced in a Transverse Mercator Projection (UTM Zone 20), using the NAD 27 Datum and Clarke 1866 Spheroid. The RadarSat image map served as the image map base for this study, as it provided coverage of the entire island.

The ETI team performed a structural analysis on an orthorectified November 24 1999 RadarSat image. Interpreted geological features were digitized into an ARC/INFO GIS format, and the data were plotted on an electrostatic plotter. All of the satellite imagery and the structural analysis were projected into a Transverse Mercator Projection (UTM Zone 14), using the NAD 27 Spheroid and Datum and printed at a scale of 1:75,000 as overlays on RadarSat imagery.

Software and Models

Most data compilation was performed on Windows 98 and 2000 platforms. Data acquisition requiring intensive tabular data input or digitizing was outsourced and may have been performed on different platforms.

The primary software for data integration was ArcView 3.2, ViewLog, PRMS, and Idrisi32. The electric logs were digitized using DigiRule's "Rat" and the associated software EarthTOOLS®. All final map products were either direct outputs or derivatives of outputs from ArcView. ViewLog™ was used for subsurface data visualization and the Precipitation Runoff Modeling System (PRMS) for recharge analysis. Idrisi32 was used for watershed delineation and georeferencing of aerial photographs. The descriptions of these programs have been abstracted from the manufacturers' promotional literature.

ArcView is an object-oriented desktop GIS produced by ESRI, developers of ArcInfo™. It was designed to run on, among other platforms, personal computers that run MS Windows or NT and on UNIX workstations. ArcView was designed for people who regularly use spatial information for decision-making.

ViewLog™ provides an integrated borehole data management and interpretation system. By tightly coupling a powerful borehole data editor with GIS style mapping and cross-section tools, ViewLog™ offers an exceptional level of visual interpretation control. The Borehole Editor was used to integrate and interpret various downhole measurements, ranging from geophysical logs to core description and core photos, and to correct, interpret, and prepare data for stratigraphic correlation, mapping, and geological modeling. The mapping/cross-section tool acted as a CAD/GIS/drawing system for interpreting and extending the borehole data. The system intrinsically supports high-level objects such as boreholes and three-dimensional surfaces and has integrated functions for gridding, geostatistical kriging, and color contouring. The key ViewLog features used include:

- Borehole Data Management
- Borehole Log Data Editor
  - Borehole Log Plotting
  - Advanced Geophysical Log Processing
- Basic Mapping and Cross Sections
- Drawing Editor Overview
  - Basic Map Editing
  - Basic Cross-Section Editing
- Integrated Data Management and Interpretation
- Open Database Structure
  - Geological Model Construction
  - Advanced GIS Mapping
  - Advanced Cross Sections and On-Section Interpretation
  - Geostatistics and Kriging
  - Ground Water Modelling

The Idrisi32™ program included the latest object-oriented development tools, including change and time series analysis, multicriteria and multiobjective decision support, uncertainty analysis, and simulation modeling. TIN interpolation, kriging, and conditional simulation were also carried out. The ETI team used Idrisi32™ tools for image processing, including restoration, enhancement, and transformation and for signature development and classification, including hard and soft classifiers and hyperspectral image classification.

The USGS Precipitation Runoff Modelling System (PRMS) is a distributed-parameter model that obtains its parameter values from available geographic coverages, including DEMs (slope, aspect), soil maps, geological maps, and land use maps. The remaining parameters must be manually calibrated. A key feature of the PRMS is the concept of the hydrologic response unit (HRU). A companion to the PRMS is the USGS's GIS Weasel, which uses a combination of interactive input by the user and predefined procedures to carry out the geographic processing and extraction of the corresponding parameters from geographic coverages to be used by the PRMS.

The Digi-Rule "Rat"™ is an all-purpose portable X-Y digitizer. The Rat was originally designed to digitize well logs for the petroleum industry, but now its uses have greatly expanded into mapping, seismic work, and chart analysis.

Software Applications. Spatial databases, described below, were the primary tools for integration of data acquired by this study. Creation and maintenance of the spatial data employs, as its main engines, the ArcView 3.2/ArcInfo™, VIEWLOG™ 3.0, Idrisi32™, and AutoCAD™ R14 software systems.

ArcView™ and ArcInfo™ were the main mapping systems of the project. These applications were used for digitization of paper maps received from various agencies, on-screen digitization of interpreted features, building of topology, and topological editing of maps. In addition, map transformations were performed in ArcInfo™. Raster images were georeferenced within ArcView™ using an extension (registerimage.avx) purchased from the Geospatial online website. Other useful extensions, downloaded from the ArcView™ website, were add, xtools, and edit tools that were used to manipulate themes and add spatial data to the tables associated with these themes. Point data, such as wells, exported from the access databases and Excel™ files were also added as themes and converted to shape files. Conversion of DEMs into contoured vector files was achieved with the Spatial Analyst component of ArcView™. The various themes and tables were manipulated within ArcView™ to produce the desired map products, and layouts were prepared for printing. This process is discussed in detailed later in the section.

The ViewLog™ system was used extensively in the project. A project was created, which incorporates raster images, ArcView™ shape files, and links to existing Microsoft™ Access databases. ViewLog™ then facilitates easy production of maps, creation and interpretation of electric log information, contouring of surfaces, and three-dimensional visualization of surfaces with well data projected from ground surface. Another important aspect was the kriging of resistivity data for individual survey lines. The krigged surfaces were displayed in three-dimensions, and correlation and anomalies can readily be observed.

Additionally, header files were created for individual wells containing information about well construction, lithology, and water quality.

ViewLog™ functionality allows shape files to be imported into a project. These files were generally used for locational and display purposes. Raster images with their world files can be geographically referenced on the map and draped over a surface providing a visual picture of the surface when viewed in three dimensions.

The most important aspect of ViewLog™ is its active link to the access databases. The main databases used were the wells database (Tdadh2O.mdb), the well production database (Well Prod Const Data.mdb), and the lithologic database (Project Wells Log Data.mdb). The common link among each of these databases is the table of wells with unique identifiers, names, and coordinates of each well.

Electric log information needed for geological interpretation was added to the database by creating a table with the appropriate fields. The electric logs digitized with the DigiRule software and exported as log ASCII (LAS) files were imported into ViewLog™ and uploaded directly to this table. A template header file of the electric log was then created in ViewLog™ displaying the curve type, depth, and value along the curve. Queries generated in the database and the header files created were then added to the data source in the ViewLog™ project, and complete cross sections were generated. These cross sections were printed and manually interpreted. Electric logs were also superimposed on scanned seismic sections and provided interpretations of these sections. These sections were then completed in ViewLog™, and plates of the interpreted cross sections were produced.

Individual electric logs were also used to count sand thickness, and these data were entered into the database. From this data, net sand thickness isopach maps were then generated. During automated contouring of the net sand thickness, the faults were interpreted from the seismic sections and used as break lines in a more rigorous kriging routine. A three-dimensional view of the top and bottom of the sand surfaces was generated and the wells were projected into the surface. This allows the investigator to "fly through" the terrain and discern high and low points and areas where channels may act as aquifers.

Production of template header files with lithology and well construction required (1) creating a library of symbols for displaying lithology, well construction, and water quality data; and (2) producing header template map files with relevant well data and the company name and logo. The lithology database tables and well production construction database tables were linked to the template header file, the appropriate symbols were plotted, and the template map file was linked as a header image to the page setup.

Another database contains the resistivity data from the geophysical surveys carried out by Earthwater Technology. These data were not linked to the previous databases. The resistivity data taken at each shot point along a geophysical line were coordinated through the shot point location on the line. The line number, distance along a line, depths, and resistivities and inverse resistivities were uploaded to the resistivity database. Each data point was uniquely identified through the line number and distance along the line. The table containing this information was added as a data source to ViewLog™. A grid covering the extent of the geophysical survey was created, and data parameters of the top and bottom surfaces of the inverse resistivity values were krigged. A cross section of the paired surface (top and bottom) was then krigged for each geophysical line. All cross sections for geophysical lines within the gridded area can be displayed in 3-D, and areas of high inverse resistivity values that indicate the presence of water bodies can be determined. Different views of a 3-D "fly-through" were saved as images.

Idrisi32™ was used to register scanned aerial photographs. This process uses coordinated points on georeferenced images of 1:25,000 topographic maps of Trinidad, and these points correspond to image points on scanned aerial photographs. The project module uses the correspondence file, which contains image coordinates and UTM coordinates, and transforms the image from an image coordinate system to the world coordinate system (UTM-20N). During this transformation, residuals of points and the root mean square error of the transformation process were given. This allows outliers to be detected and removed before the transformation was completed.

The 1:50,000 geological maps were scanned and vectorized. The island coverage was converted to a "dxf" file and imported into AutoCAD. The geology was then extracted into eight separate drawings in accordance with the 1:50,000 geological map sheets. Lines defining polygons were converted into closed polylines and exported as "dxf" files that were topologically edited in ArcInfo™. An autoLISP macro program was written for importing geological (strike and dip) data into AutoCAD. The data were added to ArcView™ through the Cad Reader extension and then converted to shape files.

The production of the map plates was done using ArcView™ 3.2. Relevant vector files and raster images were added to a view. Cartographic conventions were adopted for the display of cultural and geological features. The views were then labelled by field values from the feature attribute tables of the vector file. In some instances, the labels were individually edited and oriented for proper presentation. The symbols, line, and point types and sizes were always chosen for effective display and to accentuate the theme of the layout. Once a view was completed, the view was then inserted into a layout. A template was created to insert the views and produce the final plate. The considerations for preparing the template were to incorporate all essential elements of a finished map. Therefore, the template consists of the view with the necessary map data prepared, a north arrow, scale bar, title, customized legends, projection and datum information of the map, date it was produced, the company name and logo, as well as a plate caption.

The ArcView™ plate production groups were as follows:

1. Geology
2. Geology with interpreted subsurface structure
3. DEM with aquifer systems and surface structure
4. Production and recharge, based on aquifer systems and compartments surface structure

Topographic Maps were recreated by the team in a common coordinate system. Thirty-eight DRGs (1:25,000 scale) were obtained from LandInfo™ Inc., Greenwood Village, Colorado, for use as the project base map. These files were registered to UTM Zone 20, Clarke 1866, NAD 27, a coordinate system commonly used in the United States. Great Britain also used this system for two map series of Trinidad, but the maps were found to contain significant errors, and a third set was scanned by LandInfo. This set also had to be georeferenced with the Spatial On-line Module Register Image in order to use them in the GIS. A 1000-m grid covering the 38 sheets was created in AutoCAD™ and added to an ArcView project. The grid intersections of the AutoCAD™ drawing, and corresponding grid intersections of the DRG were used as control points to register the images. The accuracy of the process was determined by calculating the differences between the coordinates of trigonometric stations on the images (after image registration) and the actual UTM coordinates of those stations. The accuracy was ±5 m east and ±5 m north. This accuracy was only possible through this method because the topographic maps were on an orthogonal grid system, as were the images.

All DEMs were two-dimensional arrays, where each element in an array represents the average elevation of a small, four-sided patch of terrain. These small patches typically have dimensions that were given by either a fixed angular measure or a fixed length. DEMs for this project were obtained from Landinfo. These DEMs correspond to 36, 1:25,000 scale topographic quadrangle sheets and were merged into a single, 74.8-MB (USGS format), island-wide grid covering over 4700 km². This merged DEM has 3087 rows and 4087 columns (about 12.6 million cells). The geographical data were processed to enhance the geomorphology of Trinidad, permitting interpreters to identify structural and topographic features representing potential bedrock faults, fracture zones, and overdeepened bedrock valleys. Natural (i.e., noncultural) linear features called "lineaments" of more than 1 km in length were delineated from the DEM and remote-sensing imagery. Careful analysis of such lineaments by skilled image interpreters with a sound knowledge of local and regional tectonics and geology can result in identification of fractures, foliation, joints, and bedrock contacts. A new digital, GIS-compatible, and more terrain-sensitive depiction of surface watersheds was generated by processing the merged DEM to remove depressions, or pits, that hinder flow routing. The purpose of pit removal was to create an adjusted DEM in which the cells contained in depressions were raised to the lowest elevation value on the rim of the depression. Each cell in the adjusted DEM then becomes part of a least monotonically decreasing path of cells leading to an edge of the image. A path was composed of cells that were adjacent horizontally, vertically, or diagonally in the raster grid and that steadily decreased in value.

Soils Data

Soils data for the entire island were acquired from the Centre for Geospatial Studies at The University of the West Indies. Data were derived from the soils map sheets of Trinidad, input scale 1:25,000, originally defined by the Land Capability Survey and the University of West Indies in 1973. For this study, soil textures were classified as sand, sandy loam, clay, or clay loam based on the primary texture of the soil.

Subsurface Data Acquisition and Analysis

Direct sampling of rocks and sediments while drilling represented a primary source of subsurface data. However, as in most areas of the world, Trinidad's previously documented lithological descriptions from existing water wells contained imprecise, unorthodox, and unstandard-ized nomenclature such as "hard stone" and simply "clay" or "sand," and these data were of limited value for correlation and geological facies interpretation. Fortunately, most wells were also electrically logged using spontaneous potential and resistivity methods, providing a basis for geological correlations and lithological interpretations.

Where sand and gravel samples were collected for sieve analysis for designing gravel packs and selecting screen sizes, the sieve data were also used to estimate aquifer properties. This exercise was undertaken for the available historical sieve data, wherever grain size distributions yielded realistic estimates of hydraulic conductivity (K) based on the empirical estimation techniques that were used (see below). Using these methods, hydraulic conductivity (K) was also determined from sieve data from well-described sedimentary aquifer samples from test drilling during this study. Estimates of hydraulic conductivity (K) from these samples were compared with pumping test results from the same aquifer intervals. Limited pumping test data were also available for some existing water wells. A few of these have accompanying sieve analysis data that can be compared with the pumping test results.

A large dataset of air permeability, porosity, and mean grain size measurements, from mainly fine-grained sands from the subsurface Columbus Basin, was made available for study through the generosity of BP Trinidad and Tobago LLC (BPTT). These data were used to augment the hydraulic conductivity (K) estimates based on historical sieve analysis and to fill in gaps where such data were not available and are discussed in detail in the full report.

The ETI team compiled a database containing 1917 well and drill holes covering Trinidad. This information was derived from a variety of sources, including the Water and Sewerage Authority, Petrotrin, the Ministry of Energy, and National Quarries of Trinidad and Tobago. The wells were categorized as water wells, oil wells, and boreholes. The data were comprised 806 water wells, 759 boreholes, and 349 oil wells in addition to three unidentified well types.

Team geographers extracted location information for the database from the electric log records and the index sheets supplied by the data sources. The geographic data formats of data sources varied (Geodetic, Cassini, and Universal Transverse Mercator), and so ETI transformed the geodetic and Cassini coordinates to UTM coordinates with Geolog software. The accuracy of the transformation, checked using known trigonometric stations on the island, was ±1.11 m east, ±0.76 m north.

Unique identifiers derived from the full name of each well or borehole were determined for each record in the database, and these identifiers were used to link all other data in the various databases.

Aquifer Property Determination from Existing Well Data

The team analyzed all the well data to quantify water resources in Trinidad. Pumping tests inherently provide the most direct determination of these aquifer properties because the process was the quantifiable aquifer response to a known magnitude and duration of applied stress. The main advantage of determining aquifer parameters from pumping tests was that the water level response was representative of a large portion of the aquifer as opposed to samples in the immediate vicinity of the well, as was the case with sieve data. Although all new ETI-Lennox production wells were extensively test pumped, pumping test data for existing WASA wells was sparse.

Pumping Test Analysis

As most existing pumping test data were in the form of step tests, the team used alternative methods for determining aquifer parameters. The Jacob straight-line method provided aquifer parameter values and was applied to a variety of pumping test and aquifer types. Pumping test data were plotted on semi-log scale with respect to time, and a straight-line regression was performed on successive water level values at later times for a period of one log cycle. The slope of this line was used to determine transmissivity values. This method made it possible to estimate transmissivity even when a pumping rate was not maintained for an entire log cycle of time, as was the case with most historical step tests in Trinidad.

Aquifer thickness data were sparse for WASA wells with pumping test and sieve data. Because of the availability of screen thickness values for WASA wells, it was used in place of saturated aquifer thickness. Actual aquifer thickness was either not recorded through the drilling process or indeterminate in many of the existing WASA wells. The use of screen thickness for this purpose provides a common parameter source for comparison among all wells, but it yields a somewhat misleading result when converting between transmissivity and hydraulic conductivity. The assumption was made that the entire saturated thickness of an aquifer was screened and that all wells were fully penetrating. This was an idealization of actual practices, where complete penetration of a well within an aquifer was often logistically, economically, or practically unfeasible. Using screen thickness in determining hydraulic conductivity from transmissivity for a well, which was partially penetrative through the entire saturated thickness of an aquifer, yields an overestimation of hydraulic conductivity. The opposite was true when determining transmissivity from hydraulic conductivity. The use of screen thickness as aquifer thickness underestimates transmissivity.

SIEVE DATA AND ANALYSIS

Several empirical formulas can be applied to sieve data to estimate hydraulic conductivity and, therefore, transmissivity. Because of the relatively large amount of sieve data and its widespread distribution, historical WASA and new ETI-Lennox sieve data served as the basis for comparisons of aquifer parameters between aquifer systems. Resultant hydraulic conductivities were compared with conductivities derived through pumping test analysis on wells within the same aquifer system in order to quantify differences between sieve data estimates and parameters derived from pumping tests in order to gauge the suitability of estimates of aquifer parameters from sieve data.

Where existing pumping test data were available, the Jacob straight-line method was used to estimate transmissivity (T), resulting in a low estimate for transmissivity (T) compared with constant rate test determinations. Screen interval was used as a proxy for aquifer thickness in estimating hydraulic conductivity (K) due to limitations of existing data on accurately logged aquifer thickness. For example, pumping test data from the new ETI-Lennox Sangre Grande P1 production well were compared with K estimates from sieve analysis data from the same aquifer interval to calibrate methods for estimating K from sieve data for other wells where pumping test data were not available. Three empirical methods for estimating K from sieve analysis data, Driscoll, BPTT, and Sheppard, were applied and compared. The Driscoll method provided reasonable estimates of K for the widest variety of sediment size and sorting classes (Figure 53), but tended to slightly overestimate K for very fine, well-sorted sediments, where the BP equation provided a more accurate solution.

However, in certain areas of Trinidad, ETI investigators found that all three methods so overestimated (K) for very coarse sands and gravels that these techniques could not be applied to such aquifers (e.g., Northern Valley gravels and Northern Range Front gravels).

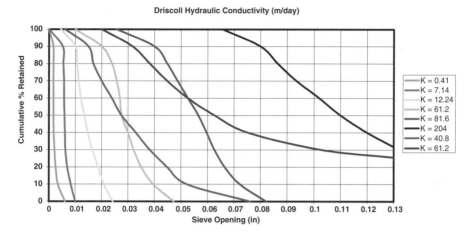

Figure 53. Driscoll estimates of hydraulic conductivity based on grain-size distribution curves.

All of the empirical formulas used to yield hydraulic conductivity values through the analysis of sieve data have limitations and advantages. A graphical comparison between the BPTT method and the Sheppard method with respect to the range of USGS sediment size classification is shown in Figure 54.

The largest hindrance on the effectiveness of the Sheppard method in characterizing a range of hydraulic conductivities was the use of median grain size (d_{50}) to represent samples. In effect, this assumption reduces the role of sorting on the hydraulic conductivity of a sedimentary unit. Although sorting was accounted for by the shape factor and exponent associated with the respective depositional environments, variation within those environments was not taken into account. Even though the median grain size may be large, the finer particles present in poorly sorted sediment can block poor throats, thus reducing hydraulic conductivity. Many of the alluvial gravels were poorly sorted with large grain sizes and clay-matrix support. Although it was less severe than with the BPTT equation, the result was overestimated hydraulic conductivities for coarse-grained sediments using the Sheppard equation (Figure 55).

Many of the sedimentary aquifers located in the central and southern portions of Trinidad were relatively fine grained and well sorted, but this was not the case in the northern regions of the country. Because of the shortcomings of the BPTT and Sheppard methods in character-

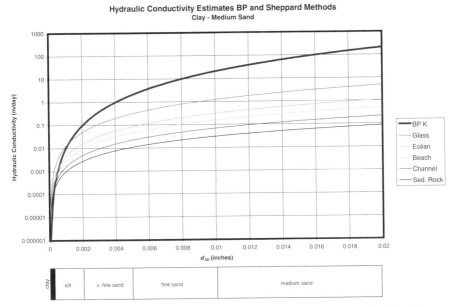

Figure 54. Plot of BPTT and Sheppard equations of median grain size (d50) of clay to medium sand.

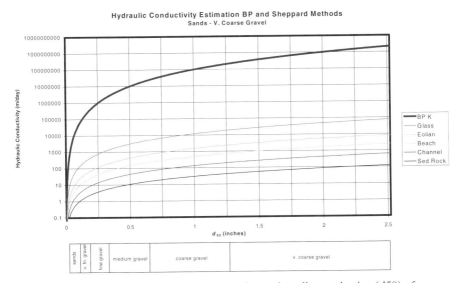

Figure 55. Plot of BPTT and Sheppard equations of median grain size (d50) of coarse sand to gravel.

TABLE 12. Results of Hydraulic Conductivity (K) Estimates from All Methods Used*

WASA ID #	Well Name	Average d_{50} (in)	Driscoll (K)	Sheppard Beach (K)	Bp (K)	Step Test (K)	Constant Rate Test (K)
1082 LL	Las Lomas #7	0.007	8.1	13.4	13.3	7.0	—
SG-1	Sangre Grande #1	0.022	24.2	81.3	2,909.2	9.9	21.42
2209TV	Tucker Valley #30	0.131	88.8	985.6	239,938	9.4	—

* All conductivities were in units of meters/day. Constant rate pumping tests were not conducted on all wells. Sieve data for SG-1 was from ETI test well SG-T5, located within 5 feet of SG-1.

izing the wide range of grain sizes and sorting properties found in the sedimentary aquifers of the country, the Driscoll method served as the basis for aquifer parameter estimation from sieve data. Although the Driscoll method tended to slightly overestimate very fine, well-sorted samples, it provided reasonable estimates of hydraulic conductivity for the widest range of samples.

Data from the three wells presented in Table 12 expose the relative quality of estimations produced by each of the three methods as compared with values derived from pumping test analysis. Table 12 also highlights the sensitivity of each solution to the median grain size (d_{50}). The average of the median grain sizes for each sample was given for the three wells as a rough basis for comparison of the general size classification of the aquifer.

These are examples of the few wells with both sieve and pumping test data. Of those shown, only Sangre Grande #1 (ETI SG-P1) has both step and constant rate pumping test data.

Sangre Grande #1 (ETI SG-P1) was constructed within an aquifer composed of predominantly fine to coarse sands, but with two distinct zones of each: an upper coarse sand and a lower fine sand. This can be seen on the plot of sieve samples along with Driscoll curves (Figure 56) Although the samples from lower intervals were fine grained and uniform, the samples from upper intervals were coarse grained and more poorly sorted. Both sands were screened in their entirety during construction of Sangre Grande #1. The Driscoll estimate of hydraulic conductivity was, by definition, an average between these two sands. This corresponds closely to pumping test data. The aforementioned relationship between step tests and constant-rate pumping tests can be seen in Sangre Grande #1 (Figure 56). The hydraulic conductivity value derived from the step test was an underestimate because of a relatively high slope inferred from the final step relative to the slope taken over one log cycle of the constant rate test. The hydraulic conductivity value estimated from the Driscoll method was comparable with that derived from the constant-rate pumping test analysis.

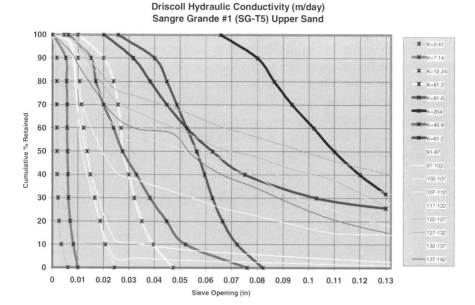

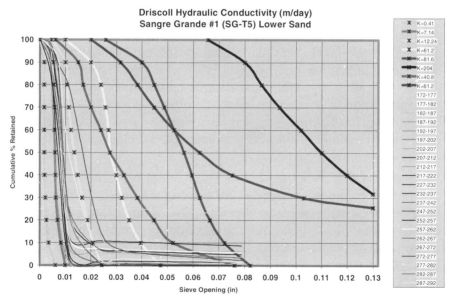

Figure 56. Size distribution curves from Sangre Grande #1 (SG-T5) plotted along with curves of known hydraulic conductivity [Driscoll, 1986]. Grain-size distribution curves of the upper sand (91'–142'). Grain-size distribution curves of the lower sand (172'–292'). (See color insert.)

Geophysical Techniques and Surveys

Borehole Geophysical Logs   Borehole geophysical logs play a critical role in evaluating the aquifer potential and physical characteristics of a penetrated sedimentary succession. Such logs include Natural Gamma Ray (Gamma), Spontaneous Self-Potential (SP), Resistivity/Conductivity, and the so-called porosity logs (Neutron, Sonic, Density). Without going into the details of tool theory and design, it may be said that the shapes of the log traces, or "log character," especially for the Gamma, SP, and Resistivity logs, can be used to interpret the vertical succession of lithologies and sedimentary depositional environments. Although it is beyond the scope of this case study to examine geophysical borehole tools in detail, a brief discussion of Gamma, SP, and resistivity logs as used in this study is included. For additional information on such techniques, the reader is referred to existing introductory texts on the subject [e.g., Asquith, 1982].

Natural Gamma Ray (Gamma)   The Gamma tool passively measures the natural gamma ray emissions of radioactive elements in sedimentary minerals. The three main substantially radioactive elements in sediments, namely, uranium, thorium, and potassium, occur most commonly in clays and shales. Quartz and arkosic sands, formation waters both saline and fresh, and hydrocarbons all elicit very little response from the Gamma tool. Glauconitic sands, on the other hand, with their high potassium and uranium content, often register a kick on the Gamma log. The Gamma tool was very sensitive and produces a high-resolution trace that can detect bedding scale variability.

Spontaneous Self Potential (SP)   The SP tool predated the Gamma tool and for many years was the mainstay of the subsurface geologist. It has a lower vertical resolution than the Gamma tool but remains useful, in that under similar borehole conditions, many lithologic units used in subregional correlation exhibit a strongly recognizable SP log-shape. The SP tool was used extensively in this study because most of the older oil wells from the Southern Basin and all of the invaluable Domoil SDH boreholes lack Gamma logs. The SP tool works by measuring the spontaneous potential of the electric current that develops across the semipermeable membrane, formed by the drilling mudcake, between the formation fluid and the drilling fluid. Currents were plotted in decreasing negative millivolt values from right to left. A "shale line" value between –20 and –35 millivolts was typical. Usually, the formation fluid was more saline than the drilling mud and formation wet sandy intervals kick to the left of the shale line on the log trace. For formation water of lower total dissolved solids (TDS) or with saline drilling mud, freshwater sands may actually kick to the right on the SP trace.

SP and Gamma logs, in conjunction with outcrop and seismic facies, were both used extensively in this study to interpret sand/shale successions and sedimentary environments in the Northern and Southern Basins.

Induction Resistivity (Resistivity)   Resistivity tools measure the resistance to an induced current flowing between two tool electrodes through the formation and formation fluids. The unit of resistivity was ohm meters squared per meter (ohm.m^2/m) or ohm meters (ohm.m). Formation resistivity was controlled by its mineralogy, pore volume and pore geometry, formation fluid composition, and ambient temperature. Shales and clays have low resistivities, and quartz and carbonates have high resistivities. Larger pores have higher resistivities than smaller pores for formation fluids of the same composition. Fresh water and hydrocarbons were both highly resistive, but saline water was conductive (low resistivity) because it contains mobile cations. Within normal geothermal gradients for sedimentary basins, resistivity decreases with increasing temperature. Trinidad basins were geothermally "cold," and for shallow, hydrogeological investigations, temperature was not an important factor in interpreting resistivity logs.

Although it may be difficult or impossible to isolate the individual effects of various factors contributing to the recorded values on a resistivity log trace, this may not be necessary for the purpose of identifying probable aquifers. This was because the most important factors that contribute to aquifer performance were all positively correlated to increased resistivity.

For example, in a coarse, clean, quartz sand containing fresh water, quartz was an insulator, clay minerals that would impede permeability while reducing formation resistance were absent, and pore volumes that increase resistivity were high; and, of course, the formation water was fresh. As porosity and permeability were positively correlated and increased clay content reduces permeability, it was clear that anything that reduces the response on a resistivity log was negatively correlated with favorable groundwater potential.

In general, a low (flat) resistivity profile opposite a strong, cylindrical, SP, or Gamma deflection indicates saline/brackish formation water.

Conversely, a high resistivity profile opposite a strong Gamma deflection and a depressed SP suggests the presence of hydrocarbons. Other log analysis tests can confirm this. However, high resistivity, opposite strong SP and Gamma deflections was an excellent indicator of fresh formation water.

Seismic Surveys

Seismic surveys generate percussive waves from specifically positioned energy sources, usually dynamite, that penetrate the Earth and are reflected and refracted back to the surface. At the surface, energy from the returning waves is recorded by an extensive array of carefully arranged geophones, and the data are electronically stored in a digital format. For two-dimensional seismic surveys (2-D), the charges and geophones are arranged so that, following processing, the seismic profile images a vertical cross section of the crust to a certain depth of resolution.

Compressional and shear waves both respond to the physical properties of the medium through which they were transmitted, in this instance, the sedimentary rock sequence. Rocks of differing density and other physical properties, such as induration or rigidity, result in a velocity interface that was registered as an acoustic impedance. Following processing, which compensates for many forms of wave divergence and removes unwanted "acoustic noise," velocity interfaces such as bedding contacts between sand and shale, faults, or rigidity contrasts between consolidated and unconsolidated rocks, show up on the seismic profiles as high-, low- or medium-impedance traces. These traces are interpreted and mapped by the geophysicist and geologist. For full treatment of the principles and application of seismic geophysics, the interested reader is referred to existing texts on the subject [e.g., Payton, 1977; Kleyn, 1982; Jenyon and Fitch, 1985].

During this study, experienced ETI team geophysicists and geologists mapped and interpreted previously unmapped horizons from a major seismic survey in the western Northern Basin and interpreted structural styles from limited seismic data from the Southern Basin. The new Northern Basin seismic interpretation and maps played a major part in our revised interpretation of the late Pleistocene deformation in Trinidad and in developing the geological model for the newly discovered Oropouche Aquifer System.

Seismic Interpretation and Mapping

The methods employed in the interpretation of the seismic data from
the Northern and Southern Basins were decided on after a careful
examination of the various qualities of seismic data as well as the
amount and reliability of other supporting information such as electric-
log data that could be used to calibrate the seismic interpretation. This
examination led to very different approaches in the type of interpretations that were attempted in the two basins.

In the Northern Basin, the seismic set consisted of 17 profiles covering 323 line km of two-dimensional data from two different surveys.
The primary data set was the Northern Basin Consortium (NBC)
survey acquired in 1996. That survey included 15 lines covering 282 km.
Two additional lines, recorded for the Ministry of Energy and National
Resources in 1984, totalled 41 km. The NBC data proved to be virtually noise free and of excellent quality. The frequency content within
the upper 1 second was dominantly greater than 100 Hz. At the relatively slow velocities in the shallow section of the Northern Basin, it
was determined that complete resolution of individual beds thicker
than 5 m (15 ft) was possible. ETI decided to attempt a seismic sequence
style of interpretation because the data were suitable for the delineation of bedding geometry. One drawback to this approach was the
orientation of the survey in which 13 of the profiles were acquired in a
regional dip direction and only two strike lines were included for line
ties. However, the dip lines were spaced at approximately 2 km, and a
considerable amount of well control (electric-logs) was available for
correlation.

The team began interpretation by identifying all of the major faults
and unconformities that appeared on the sections and by correlating
these in a "loop-tie" manner. Then, using ViewLog™ software, all wells
that fell within 100 m of each of the seismic profiles were projected into
the two-dimensional line of profile. At the same time, all lithologies and
formation tops that appeared on the surface geological maps were also
plotted onto the seismic sections. This sometimes resulted in as many
as 20 well-ties into a single seismic profile along with three to four
surface ties. By correlating and identifying the various formations on
the e-logs and from outcrop, it was possible to then place the seismic
events into a chronostratigraphic order. This often became an interactive process because it was soon discovered that certain historical
formation correlations did not conform to the structural picture that
was evident on the seismic profiles. An example of the result is five
sequences newly identified in the Talparo Formation. Time structure

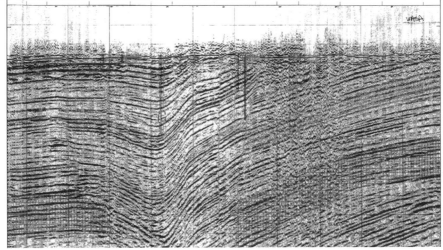

Figure 57. Portion of SBC Seismic Line 6 shows interpretation of Talparo sequences overlaid on seismic reflectors.

maps at the bases of the sequences were generated and loaded into the ViewLog™ package for further analysis (Figure 57).

The data quality in the Southern Basin proved to be much more problematic. The sole data set used in that area was approximately 2800-km two-dimensional data that were acquired in 1990 and 1991 by Exxon Trinidad Limited for the Southern Basin Consortium (SBC). These data included 39 profiles oriented in a grid that measures approximately 2 km × 7 km, with the closer spacing being perpendicular to structural strike. The frequency content of the data was relatively low (i.e., less than 40 hz), and the data were often saturated with noise, especially in the anticlinal areas.

Although abundant surface contacts were available for correlation, the poor data quality made such methods questionable. Although a number of water and oil and gas wells have been drilled in the basin, that control was often clustered, leaving e-log correlations in question. We therefore decided to rely on the SBC data to confirm locations and depths of synclines and to identify the structural styles and fault orientations within the basin.

It should be noted that, except for the two seismic sections in the Northern Basin acquired in 1984, ETI was provided with only the shallowest 1 second of seismic data in both basins. Although best efforts were made to correctly interpret the structural styles and bedding geometries, further efforts that include the deeper data set could result in alternative interpretations.

Controlled Source Audio-Frequency Magnetotellurics (CSAMT)

Controlled Source Audio-Frequency Magnetotellurics (CSAMT) is a high-resolution electromagnetic sounding technique that uses a fixed, grounded dipole or horizontal loop as an artificial signal source. CSAMT is similar to the Natural Source Magnetotellurics (MT) and Audio-Frequency Magnetotellurics (AMT) techniques, with the main difference being the use of an artificial signal source for CSAMT. The source provides a stable signal, resulting in higher precision and faster measurements than are usually obtainable with natural-source measurements in the same spectral band. However, the controlled source can also complicate interpretation by adding source effects and by placing certain logistical restrictions on the survey. In most practical field situations, these drawbacks are not serious and the method has proven particularly effective in mapping the Earth's crust in the range of 20 to 2000 m.

Geological Mapping

There is no doubt that the occurrence of groundwater sources in Trinidad is a creature of geology. Given Trinidad's high rainfall, near-surface geology represents the primary controls on meteoric water infiltration and percolation, and deeper geology defines underground water storage and transmission. Accurate geological mapping was therefore fundamental to the success of the new groundwater assessment. ETI's GIS-integrated convergent analysis of spatial data standardized all map data on a digital ArcView platform and enabled the team to work with an extremely user-friendly and interactive working basemap. The geology of Trinidad, far more so than many places in the world and certainly Tobago, has been intensely studied for the last 100 years. This has been mainly because, in one way or another, of the commercial presence of onshore and offshore hydrocarbons associated with the island. During the first half of the century, Trinidad's oil formed a strategic commodity for the British Colonial Government and the Royal Admiralty during World Wars I and II. In the years prior to seismic geophysics, surface geological mapping provided the main means of hydrocarbon prospecting. It was during this time, prior to 1960, that most of the intensive surface field mapping of the island was undertaken.

Trinidad's complex geology has always interested geologists. Without the imperative of a hydrocarbon-based economy, the intensive histori- cal investment spent on studying the island's geology would not have been feasible. With the development of Plate Tectonic Theory in the 1960s, geologists worldwide found a new reason for studying Caribbean geology. Trinidad's position within and at the leading edge of the Caribbean/South America Plate boundary zone now makes the study of its geology immensely interesting to structural geologists and basin analysts with little direct interest in hydrocarbon systems.

The value of having access to prior geological studies for an island as complex as Trinidad cannot be overstated. In light of previous work, this project could not have been meaningfully undertaken without a full evaluation of the existing canon of geological knowledge for the area. This presented particular challenges for the project team as no prior digital geological map of the island was available at the start of the project and much of the key structural and stratigraphic studies remained unpublished in industry files and were not available to the team. Other key subsurface data, such as NBC and SBC seismic and some oil well e-logs, were made available and have proven invaluable for our revised interpretation of some aspects of the geology. In October 2001, the availability of Kugler's long awaited Tertiary text for his Treatise on the Geology of Trinidad was a late-stage trove of data and insights into the development of the local stratigraphy. Every effort has been made to incorporate the relevant information from this volume into the study.

The foundation and context of ETI's surface and subsurface mapping work on the island is contained in a thorough knowledge of the history of prior studies. Much of the information was available in the authoritative *History of Trinidad Oil* by George E. Higgins [1996]. The Kugler, Persad and Saunders maps were discussed in the context of the revisions and preferences of the present digital geological map produced for this project.

HISTORY OF GEOLOGICAL MAPPING IN TRINIDAD

Modern geological mapping of the island began in earnest with the publication in 1860 of Wall and Sawkins' seminal British Geological Survey memoir on the geology of Trinidad [Wall and Sawkins, 1860]. These authors delineated and mapped major lithologic series, includ- ing metamorphic rocks of the Northern Range, Northern Basin allu- vium, the Central Range limestones, argillites at San Fernando and the

Moruga deltaics, as well as asphalt occurrences and seeps at and around La Brea. In 1959, a geological map of Trinidad was published by Kugler.

Twentieth-century exploration geology in Trinidad started in earnest with the detailed, often 1:10,000 scale, field mapping of Government Geologist E. H. Cunningham Craig [1904, 1905a,b,c, 1906, 1907a,b,c]. His maps and reports of the Southern Basin anticlinal structures remain valid primary data sources of surface geology and were included in Kugler's compilation map.

Starting in 1913, a Shell subsidiary named United British West Indies Petroleum Syndicate (UBWIPS) put a party of geologists into the field under the direction of B. Macrorie and John "Bullie" Bullbrook. They were charged with surveying areas from Matura south to Moruga and then west to Cedros. Trinidad Leaseholds Limited (TLL), the company that played the most important role to date in developing geological understanding of onshore Trinidad, was formed in the same year. With TLL came the arrival of the Swiss geological contingent to Trinidad: Dr. Fortunat Zyndel, Dr. A. Tobler, and, later, Dr. Hans Gottfried Kugler. This team, together with the Malayan surveyor A.K. Mas Bakal and Austrian geologist Dr. L. Sommermier, mapped extensively in the Guayaguaywere area at the same time that Bullbrook's Shell party was active in the field. Zyndel later studied and mapped the petroleum geology of the Guapo—Forest Reserve—Siparia area.

Hans Kugler was an early advocate of the application of planktonic foraminifera to Trinidad stratigraphic studies. When H.M. Bolli and P. Brönnimann joined TLL, in 1945, Kugler encouraged them to undertake systematic taxonomic studies of the planktonic foraminifera of Trinidad. Building on the work of Renz and Stainforth, they developed the modern planktonic zonation for Cretaceous to Miocene units that was published by Bolli in a series of seminal papers [Bolli, 1957a,b,c]. By establishing the chronostratigraphic relationships of the shale-prone and structurally complex lower Tertiary sections of the Southern Basin, this work had tremendous economic and scientific impact. It was not an overstatement to say that this work formed an indispensable basis for the contemporaneous compilation, by Kugler, of the Trinidad geological map.

By the early 1950s, a large number of lithostratigraphic units were in informal use among the oil companies and a great many more that had been used were now obsolete. The informal nature of the stratigraphic nomenclature caused considerable confusion and miscommunication among professional geoscientists. This was especially so as most of the reports setting out stratigraphic concepts, and the substantiating data, were unpublished and held in company files. In fairness, the informal and unstandardized state of stratigraphic terminology was a worldwide problem at the time.

In 1955, in response to the problem and as part of a worldwide effort to produce an international stratigraphic lexicon, Kugler organized a meeting of the leading geologists working in Trinidad, with an aim to stabilizing (if not standardizing) the local nomenclature. This conference has come to be known in Trinidad geological lore as "The Meeting of Pointe-a-Pierre" [Saunders and Bolli, 1968]. The results of these discussions were compiled by Kugler for the Trinidad section in the International Stratigraphic Lexicon [Kugler, 1956].

Between 1951 and 1952, Dr. Hans Suter of TLL published an updated and expanded treatment of the geology of Trinidad, which, with the Lexicon, provided a published state-of-the-art for the time. A revised edition, with an important appendix by George Higgins, was published in 1960. Kugler, Suter, and Higgins were all at TLL in the 1950s, as were Rohr, Bolli, Barr, Bower, and Saunders. It was during this time, in parallel with the Lexicon project, that Kugler was compiling and working on his "Treatise on the Geology of Trinidad."

A major part of the Treatise project was the compilation by Kugler of an island-wide geological map. This was begun at 1:25,000 scale, but the progress was slow. Following the Texaco buy-out of TLL, the map 1:25,000 sheets were condensed and completed at 1:50,000 scale. These fine maps were hand drafted at the Texaco Geological Drawing Office in Pointe-a-Pierre. From these, a 1:100,000 map with seven north-south sections was produced in Pointe-a-Pierre and printed in Zurich in 1961. This colored 1:100,000 map instantly became the accepted geological map of Trinidad and has withstood the test of time in very many respects. Chief among these was the ground-truth location and description of lithologic types, the position of faults, synclines, and anticlines, and the stratigraphic relationships among formations.

Between 1959 and 1977, following his departure from Trinidad, Hans Kugler compiled the detailed stratigraphic text for the Tertiary formations covered on his geological map. This text was cross-referenced to a series of 32 spot maps and sections, at scales of 1:10,000 or smaller, that had been prepared in Pointe-a-Pierre for the Treatise project. Publication of the Tertiary text and the spot maps was delayed following the death of Kugler in 1986. In 1996, the spot maps and sections were published posthumously by the Natural History Museum in Basel, Switzerland. In 2001, the Tertiary text of the Kugler Treatise was published by the museum. Kugler never completed the text for the Mesozoic formations. Therefore, the Treatise project comprises four parts: The Stratigraphic Lexicon [Kugler, 1956], the 1:100,000

Geological Map with seven sections [Kugler, 1961], the spot maps and sections of type localities [Natural History Museum, Basel, 1996], and the Paleocene to Holocene formations, or Tertiary text [Kugler, 2001].

The Saunders Map

The modern era of geological work in Trinidad was characterized by the triple technological revolutions of Plate Tectonics, powerful computers, and seismic sequence stratigraphy. Following the nationalization of the onshore industry in the late 1970s, a cadre of Trinidad-born geologists began to experiment with plate tectonic ideas in the Trinidad context. The difficulty of acquiring quality seismic data in the onshore Southern Basin hampered this effort, but some pivotal contributions were achieved [Persad, 1984; Williams, 1986; Tyson, 1989; Payne, 1991].

In the mid 1980s, a project to revise and update the Kugler map was initiated by the local geological community. The project was supported by Amoco Trinidad, which provided funding and technical assistance, and was overseen by the Ministry of Energy. John Saunders, working closely with Bruce Eggertson of Amoco, edited the compilation, which was published in 1998. The Saunders map, as it was known, was envisioned as a digital project, with multiple layers of geological, geographical, and cultural information. As a result, the printed map does not contain all of the data associated with the project. The digital product was completed by BP Trinidad and Tobago LLC (BPTT) and officially jointly launched with the Ministry of Energy and Energy Industries in October 2001. The MOE plans to market this product to the geological community, but at the time of this writing, the details have not been finalized.

The Saunders map draws heavily on the Kugler 1:100,000 map as a base but contains some key differences. Geodetically, the Saunders map was presented in UTM projection, and geologically, there were three main differences between the two maps, as described by Saunders in the notes that accompany the map: (1) The Northern Range contains significant new interpretations including new formations in the center and northwest based on the work of Algar [1993] in the northeast by Potter [1973]; (2) the age relationships of the upper Neogene Southern Basin formations from west to east were significantly altered based on foraminiferal and sequence stratigraphic interpretations of coastal outcrops by Carr-Brown and King [1996], and (3) the Saunders reintroduced the term *Concord Formation* for some of the Central Range Limestones based on limited foraminiferal data and extrapolation from outcrop data.

The Saunders map project relied on the 1:100,000 Kuglar maps and did not have the benefit of either the SBC or NBC seismic data to help with revising the surface geology. Nor did it make use of the extensive Domoil Structural Borehole (SDH) data that were used so extensively in the ETI study and that also used Kugler's 1:50,000 maps, containing considerably more surface detail than the 1:100,000 map.

ETI GEOLOGICAL MAPPING

The starting point for geological studies was the 1:50,000 scale Kugler surface geological map (Sheets A–H). The map was digitized, vectorized to closed polygons in an ArcInfo™ GIS format, colorized, and transformed from the original Cassini projection to the Universal Transverse Mercator (UTM) system (see metadata Appendix for details, vol. IV this report). From there, while keeping digital copies of the original map components, areas of the map were reinterpreted and revised, based on new data and reinterpretations of existing data. New interpretations (maps) were saved as separate coverages, allowing easy revision and preventing the loss of earlier perspectives. In effect, the ETI digital map became a living document honoring and validating the extraordinary field efforts of Trinidad's pioneering geologists while transitioning into the digital future of remote sensing and four-dimensional subsurface imaging.

Drilling Technologies

History of WASA Well Drilling Technologies   Traditionally, the mud rotary well drilling method was used to drill WASA's sand and gravel water wells. The mud rotary method has a number of shortcomings, including poor formation sample reliability, the inability to test water quality while drilling, and the inability to estimate water flow amounts while drilling.

Introduction of Dual Rotary Drilling Technologies   Accurate formation sampling can best be accomplished in fine sand aquifers using the dual rotary direct air, dual rotary water flush, and dual rotary reverse circulation methods. All three drilling methods allow for the collection of unadulterated formation samples. The dual rotary direct air drilling method also allows for accurate interval sampling for formation water quality and interval flow rate determination. All of the dual rotary methods allow for the simultaneous advancement of the drill string and the casing so that only a small portion of the formation was exposed for sampling. Dual rotary methods can also be used to effect in coarse-grained (gravelly) formations, although the rate of penetration may decline through heavy gravels. In gravel formations, the dual rotary method was best combined with an air hammer, rather than a rotary roller bit, to break-up and extract coarse-grained materials from the hole. Perhaps the biggest advantage of the dual rotary method over the mud rotary method was that the water bearing formation was not plugged with clay during the drilling process, thus reducing extensive well development.

In fractured bedrock and solutioned aquifers, the dual rotary method allows for drilling to proceed in very difficult caving formations. The metamorphic rocks of Trinidad were too hard for the mud rotary method to be effective. Conventional air rotary methods will work. However, most of the metamorphic rocks were highly layered and were subject to caving and collapse, resulting in the inability to drill deeper in the well and in the possibility of a drill bit becoming trapped in the well. As with the dual rotary method the casing can follow the bit, caving formations can be easily penetrated, and in very difficult wells, the well can be telescoped with a smaller diameter casing.

Optimum Spacing of Wells   In order to determine the optimum spacing between wells, many aquifer and well characteristics must be considered. Generally, a determination of the amount of interference between wells requires the installation of pumping and observation wells coupled with the analysis of long-term pumping test results. There were several methods of estimating the optimum distance for siting production wells to minimize well interference. Well interference was caused by the expansion of the drawdown cone laterally such that the water level of an adjacent well was lowered. In confined, highly productive aquifers, this impact can be felt several kilometers away. The methods include the application of a groundwater computer model, an analytical approach using hydraulic equations, and a recharge approach that relates recharge to expected well performance.

Groundwater Model   The purpose of a groundwater model was to assist in managing an aquifer. Specifically, a model can be used to predict the effects of long-term pumping of a well at a specified rate, or to determine how much interference might be expected among wells if one or more additional production wells were installed.

The most basic function of a groundwater flow model was to simulate the shape of the water table or the potentiometric surface in an aquifer, under a specified set of conditions. In a finite difference flow model, such as was used to model the Sangre Grande Aquifer, an orthogonal grid was superimposed on the region to be modeled, or model domain. The domain can then be seen as a collection of relatively small cells organized in layers, columns, and rows. Each cell represents a small volume of the region that makes up the model domain.

Every cell was assigned a value for several relevant variables, including hydraulic conductivity, storage, and any external variables that have an influence on hydraulic head at that point in the system. By "external variables," we mean things like a river that could contribute recharge to an underlying sand layer, or a well from which water was being pumped. In modeling parlance, these external factors were referred to as boundary conditions.

For each cell, the computer sets up an equation that was a numerical approximation of a differential equation governing groundwater flow. This equation describes the average head in the cell as a function of the head in surrounding cells and any external variables. One equation was created for each cell in the model domain, and the complete collection of these equations can be considered a matrix. When the computer model was run, it generated a simultaneous solution to the complete set of equations. The solution consists of an array of hydraulic heads, one for each cell in the model domain.

The ability to make predictions with models was the main reason they were created in the first place. Examples of predictive uses would be (1) to show the results of pumping a well at a different rate than that at which it has been tested; (2) to show the impact of placing additional wells in an aquifer; and (3) to predict the influence of seasonal or climatic changes on water levels.

Predictive models accurately based on multiple sets of calibration data were favored, and model predictions based on aquifer conditions similar to those to which the model was calibrated were found to be more accurate than those based on substantially different conditions. When setting up predictive scenarios, analysts confirmed that the elements of each selected scenario was subjectively consistent with the underlying conceptual model.

Detailed results of a groundwater model run for the Sangre Grande sands underlying the Oropouche Monocline on ETI's production well Sangre Grande 1 allowed ETI to predict with confidence that an additional well could be installed in that aquifer. Subsequent drilling and pumping tests confirmed the model's prediction.

Another approach to determine the optimum spacing of wells in a compartment was to first determine the number of wells required to extract all of the potential recharge by dividing the potential recharge by the average well yield. Then take the length of the long axis of the recharge area (exposed outcrop) for dipping formations or the total area exposed to recharge for flat lying aquifers and divide by the number of wells necessary to extract all of the recharge. The result was the optimum distance between wells. This was a more practical approach for water resource planning than estimating the zone of influence using hydraulic methods. An example of this approach for the Sum Sum sand outcrop surface area, with the total length of each compartment within the Couva-Claxton Bay Monocline, Carapichaima Monocline, and Longdenville-Las Lomas Monocline Aquifer Systems and the recharge to each of the compartments shown in Table 13.

The separation distance was a function of the minimum, maximum, and average well yields, which were assumed to be 2.3, 4.6, and 3.5 ML/ day. The number of wells was rounded up.

As was shown, there was a big difference between the minimum and maximum number of wells required to extract the total groundwater recharge. As the cost per well was nearly the same, the development cost for relatively low yield wells was substantially higher than optimally located wells because a much larger number of wells will be required to develop the same amount of water. Similarly, the operation and maintenance costs would be significantly higher because of the greater number of pumps, chlorination facilities, and increased number of man-hours required to maintain the well facilities.

As is shown there is a big difference between the minimum and maximum number of wells required to extract the total groundwater recharge. As the cost per well is nearly the same, the development cost for relatively low-yield wells is substantially higher than optimally located wells because a much larger number of wells will be required to develop the same amount of water. Similarly, the operation and maintenance costs would be significantly higher because of the greater number of pumps, chlorination facilities, and increased number of man-hours required to maintain the well facilities.

A detailed exploration program identifying the optimum location of high-yield production wells, combined with drilling technologies that maximize well efficiency reduced water development and production costs across the board. Trinidad's groundwater sources before the megawatersheds project had inefficient, closely spaced wells resulting in very high operation and maintenance costs. These costs are

TABLE 13. Optimum Spacing of Wells Using the Recharge Method

Compartment	Length of Long Axis	Recharge (ML/day)	Separation Distance in Kilometres			Number of Wells		
			Minimum	Maximum	Averag	Minimum	Maximum	Average
CCBM 1	4,800	3.27	2.4	4.8	4.8	2	1	1
CCBM 2	1,600	0.57	1.6	1.6	1.6	1	1	1
CCBM 3	2,400	1.58	2.4	2.4	2.4	1	1	1
CM 1	3,300	3.53	1.7	3.3	1.7	2	1	2
CM 2	4,000	2.96	2	4	4	2	1	1
CM 3	4,500	5.41	1.5	2.3	2.3	3	2	2
LLM1 1	500	1.59	0.5	0.5	0.5	1	1	1
LLM2	2,500	0.41	2.5	2.5	2.5	1	1	1
LLM3	10,700	15.61	1.5	2.7	2.1	7	4	5
		Estimated well yields (ML/day)				2.3	4.6	3.5

significantly lower in the 15 imgd produced by this project, with the proper location and spacing of production wells.

HYDROGEOLOGICAL ASSESSMENT AND RESULTS

Introduction

The groundwater resources in Trinidad were assessed by documenting and analyzing meteorological, geological, and hydrological processes controlling water balance, estimating groundwater recharge, and determining the amount of groundwater available for safe, sustainable withdrawal. Using a novel combination of satellite remote sensing, GIS, meteorological, and hydrological modeling, this ETI assessment quantified the water balance in 11 watersheds in Trinidad. As it was impossible to study every watershed within the scope of the project, these watersheds were selected based on their geographic distribution, geological representation, and quality of data and then applied to the rest of Trinidad in a comparative analysis of watershed attributes. The technical rationale was that these combined calibration watersheds were representative of all of the various physiographic areas and geologic terrains in Trinidad, permitting, by comparative analyses, accurate calibration of other areas with poor or no runoff data. Rainfall-recharge relationships were derived from these calibration watersheds for the rest of the island based on comparative geology and physiography, resulting in average daily recharge calculations to all aquifers in Trinidad.

ETI's working conceptual model in Trinidad used geology as the principal regulator of groundwater environments. Virtually all of Trinidad's aquifers share a tectonic genesis. Regional tectonics have induced wrench and normal faulting with significant displacement in consolidated, semiconsolidated, and unconsolidated formations throughout the island. This tectonism has had a profound effect on hydrogeological processes, from topographically controlled precipitation to the evolution of megawatersheds in Northern Range bedrock and aquifer systems in North, Central, and Southern Basins' sediments.

Evaporation, precipitation, and recharge were calculated using 12 years of data in 11 "calibration watersheds" that appeared to represent the full spectrum of hydrometeorological and geological environments in Trinidad. Daily estimates of precipitation and evapotranspiration for the period 1985 to 1997 at a grid resolution of 4 by 4 km was determined by using the proprietary CROPCAST® program (Section 3.5.2), which employs existing data and scientist-interpreted image-data from orbiting satellites to fill in gaps between recording climatological stations. To fulfill this goal, precipitation data, other meteorological data, and satellite data were integrated with soils and land cover data in a hydrometeorological model.

To estimate the natural recharge to these watersheds, the hydrologic processes that lead to runoff from precipitation were modeled using a rainfall-runoff GIS model (USGS's PRMS model) that uses topography, land use, and soils data as input. This model was calibrated inputting the 12 years of daily precipitation and potential evapotranspiration data generated by CROPCAST®. The model-simulated runoff was then compared with historical runoff records provided by the WRA. Once a satisfactory calibration was obtained, a data series was produced containing daily groundwater recharge to each of the modeling units in each watershed for the 12-year period.

Of the original 11 calibration watersheds, two were eliminated during the analysis because of redundancy or data quality. For each of the remaining nine calibration watersheds, polynomial relationships between monthly precipitation and recharge were developed. The geologies and physiographies of the calibrated watersheds were compared with those of aquifer systems and megawatersheds to determine the appropriate equation for recharge calculations. Average daily precipitation for a wet and a dry year of the primary and secondary recharge zones of aquifer systems and megawatersheds was used with the appropriate equation to determine recharge. Calibration watersheds chosen for calibration in this analysis include F2-1 Cunapo, F1-2 Matelot, F2-5 Matura, F4-1 Ortoire, F5-3 Inniss, F6-1 Guapo, F8-1 Couva Caroni Limited Weir, F9-6 Guanapo I, and F9-12 Cunupia.

The topography and elevation of calibration watersheds varies from 1500–3000 m in the steep and mountainous Northern Range to 50–300 m in the rolling hills of central and southern Trinidad and 5–50 m in the flat basins of the northern, central, and southern part of Trinidad. The surface areas of the calibration watersheds ranges from 23 km^2 to 82 km^2. The geology of the three Northern Range calibration watersheds consists of phyllites (some calcareous, creating favorable groundwater environments), quartz, and carbonate schists and slate, with metasandstone and karstified metalimestones representing major aquifers within megawatersheds. The geology of the remaining six calibration watersheds is representative of groundwater environments associated with alluvial systems and tectonically evolved synclines and monoclines.

The percentage of precipitation contributing to groundwater recharge under nonpumping conditions in the calibration watersheds ranges from 2% to 35%. The lowest recharge (2%) was in the clay-dominated recharge zone of Cunapo (F2-1), and the highest (35%) was in the fractured carbonate-rich metamorphic rocks of Matelot (F1-2).

In the Northern Range, the groundwater storage and transmission zones comprising megawatersheds and fractured bedrock aquifer systems are recharged from rainfall via (1) direct percolation into outcropping fractured rocks (especially Karst), (2) percolation through coarse, shallow soils and gravels and (3) percolation from streambed alluvium weathered bedrock into large through-going geologic structures and foliation planes (secondary recharge zones). Potential groundwater recharge in the Northern range exceeds 990 ML/day (218 imgd).

Groundwater recharge in the Northern Range foothills occurs primarily through direct infiltration of precipitation through surface gravels and soils or through subsurface groundwater flow from fractured/weathered bedrock that is in hydraulic communication with the gravels and through bottom sediments of streams as they cross over exposed gravel deposits.

Groundwater recharge to the varied sedimentary aquifer systems comprising the groundwater sources outside the Northern Range occurs through direct infiltration of precipitation into the exposed outcrop and through alluvial sediments along stream valleys that cross over exposed outcrop areas of aquifers. The recharge to the bedrock and alluvial aquifers of Trinidad is summarized in Chapter 5. Northern Basin and Central Range sediments receive sufficient groundwater recharge to represent groundwater resources of approximately 160 ML/day (35 imgd) and 60 ML/day (13 imgd), respectively. Southern Basin aquifer systems represent another 290 ML/day (64 imgd).

PRMS Model—Discussion

The hydrologic model used for the modeling of the Trinidad watersheds was the USGS Precipitation Runoff Modelling System (PRMS), which was considered the most flexible available and is generally realistic for application in sedimentary systems dominating the island. This methodology estimates groundwater recharge assuming static, rather than pumping, conditions that are reasonably representative of aquifers exhibiting medium to high porosities and low to medium permeabilities with limited access to active recharge (i.e., surface outcrop). However, this assumption does not adequately address static conditions and cannot predict additional induced recharge to deep groundwater systems under pumping conditions in fractured bedrock areas.

The greatest rainfall occurs in the Northern Range of Trinidad, where two-thirds of Trinidad's groundwater resides. The bedrock here has high surface and subsurface permeabilities, allowing increased recharge in areas where a groundwater deficit due to pumping is created. Therefore, the groundwater recharge estimates contained in this report are considered by the authors to be representative of static and pumping conditions in the island's sedimentary aquifers, but conservative (i.e., low) for Northern Range fractured bedrock aquifers under pumping conditions.

Specifically, in the PRMS model, after subtracting evapotranspiration, the remaining Earth-bound part of the water cycle is described in terms of hydrographs with three components, "surface runoff," "subsurface flow," and "base flow" (Figure 58). When these components are added to evapotranspiration and fail to balance with precipitation, unaccounted-for remaining water is relegated to another, catch-all term, "groundwater sink." The resulting hydrographs of water withdrawals in the calibration watersheds are therefore reasonably representative of natural equilibrium (i.e., static) conditions in Trinidad's slow-flow sedimentary aquifer environments and less so in the conductive Northern Range bedrock aquifers.

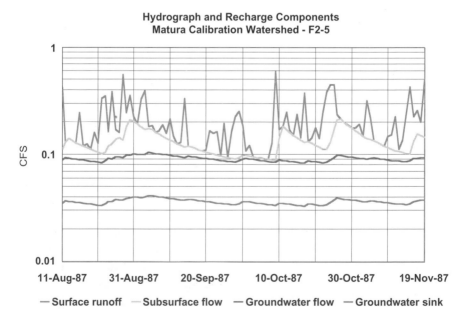

Figure 58. Hydrograph for storm events for the Matura watershed illustrate the hydrograph and recharge components. (See color insert.)

In the mountains, heavy overland flows from intense rainfall events over gentle ridge slopes might induce larger quantities of water into underlying conductive fractures, reducing runoff and increasing base flow. Megawatersheds and fractured bedrock aquifer systems found in the Northern Range also represent favorable zones for submarine groundwater discharge (SGD), which requires specialized techniques to quantify. SGD studies carried out using chemical sampling techniques off the Oronoco River by Moore [1996], off Florida's coast by Cable et al. [1996], and the Southeastern United States by Shaw et al. [1998], found that SGD was equivalent to 40% or more of surface runoff. Other independent studies by Basu et al. [2001] found that SGD into the Bay of Bengal was equivalent to approximately 20% of the combined flows of the Ganges and Brahmaputra rivers. Conclusions of these and many other recent SGD studies using state-of-the-art technologies indicate that much more rainfall is making its way into deep groundwater transmission systems (i.e., megawatersheds) than current hydrographs record. With the discovery of SGD-compatible megawatersheds and aquifer systems in Trinidad and Tobago, future studies of SGD could add substantially to the known sustainable groundwater bank.

Therefore, using the best analytical approach available, the results may underestimate, perhaps significantly, the groundwater recharge potential, under pumping conditions, of high-permeability aquifers with high static water tables and high rainfall. Long-term pumping from deep wells strategically located in these megawatersheds and aquifer systems could substantially increase safe yields by inducing more recharge from active rainfall and cycling more groundwater and subsurface flows through the deep aquifers. For example, the hydrograph for the Matura watershed (Figure 58) quantifies the three flow components based on static conditions. In this highly permeable watershed, "surface runoff," "subsurface flow," and "base flow" are quite possibly hydraulically continuous, and deep wells pumping large enough quantities of water from the deepest part of the "base flow" zone would eventually capture part of the "subsurface flow," which would in turn lower the water table and capture more water from surface runoff. The subsurface flow portion of the Matura calibration watershed hydrograph is equal to 13% of the 12-year average precipitation. Added to the 33% of precipitation that was calculated as recharge from PRMS, the total potential recharge to the groundwater system is equal to 46% of precipitation.

Under these favorable conditions, long-term pumping from deep bedrock wells could induce substantial additional recharge, effectively increasing safe, sustainable aquifer yields. Groundwater that normally would be lost to subsurface flow, thereby adding to the flooding threat, would both increase the annual safe yield by pumping and benefit downstream inhabitants. For example, if the groundwater reservoir had a deficit because of pumping, then during frequent mountain storm rainfall events, when 100 mm or more may fall in a few hours, more runoff would be captured, thereby increasing the net amount of water available for withdrawal and reducing the energy levels of destructive storm runoff. The Northern Range groundwater systems collectively receive the highest rainfall in Trinidad, with estimates of potential groundwater recharge, under nonpumping static conditions, exceeding 990 ML/day (218 imgd).

DISCOVERIES OF MEGAWATERSHEDS AND AQUIFER SYSTEMS

Summary

The Earthwater Technology team has identified and classified four types of major groundwater resources on the island of Trinidad that, collectively, have the potential to provide a sustainable yield of over 1500 million liters per day (ML/day), or 330 million imgd of groundwater, most of which is either naturally potable or treatable using readily available technologies at minimal costs.

According to WASA records, approximately 236 ML/day (52 imgd) of groundwater was extracted in Trinidad in 2001. The most recent water resources report prior to this ETI report was completed in 1999 by DHV Consultants, who compiled, evaluated, and summarized all earlier works on the subject and concluded that the safe yield for Trinidad groundwater production was 314 ML/day (69 imgd). Therefore, according to the DHV study, only 77 ML/day (17 imgd) of Trinidad's groundwater remained untapped in 2001.

The key geological factors involved in the formation of Trinidad's aquifers and the scientific basis for aquifer classification are described below, along with tables listing known aquifer parameters, range of recharge, and current withdrawal rates to each aquifer system. Table 14 provides a summary of recharge and existing production in each of the megawatersheds and aquifer systems in Trinidad.

The treatment of the metamorphic and sedimentary terranes in ETI's full report is similar in layout. First, the main concepts describing the hydrogeology of fracture (secondary) porosity aquifers and sedimentary (primary) porosity aquifers are presented. For example, the concepts of fractured bedrock aquifers and megawatersheds are introduced for the Northern Range metamorphics, and Aquifer Quality Sediments (AQS) and AQS Tracts, sequence stratigraphy, basin analysis and structurally delineated aquifer systems are introduced for the sediments comprising the remainder of the island. Secondly, the fracture systems (metamorphic) and AQS tracts (sediments) are described in detail. Thirdly, for each terrain, the megawatersheds (Northern Range) and aquifer systems (sedimentary province) are systematically treated in terms of their map delineation, hydrogeologic model, recharge assessment, current production, and estimated resource potential.

The following paragraphs summarize highlights and conclusions of this hydrogeological assessment, first for the Northern Range metamorphic megawatersheds and then for the sedimentary aquifer systems. Groundwater resources potential and other key statistics for the most important systems are presented in Table 14.

Intensive studies of Northern Range metamorphic rocks performed during this project conclude that the physical properties of the province are consistent with the formation of fractured bedrock aquifers and megawatersheds. Subsequent drilling and test-pumping of bedrock wells in bedrock aquifers and megawatersheds supported these conclusions, indicating the potential for further development of more than 931 ML/day (205 imgd) of untapped groundwater, with individual well yields exceeding 9 ML/day (2 imgd) achievable using appropriate technologies.

Within the remaining, sedimentary, part of the island, AQS tracts and aquifer systems provide the primary terms of reference for discussing the subsurface hydrogeology. In this study, 19 aquifer systems, including 12 AQS Tracts have been identified and mapped: eight aquifer systems in the Northern Basin (five AQS Tracts), with 37 ML/day (8 imgd) of combined, untapped groundwater potential, nine in the Southern Basin (five AQS Tracts) with 249 ML/day (55 imgd) of combined, untapped groundwater potential, and two in the Central Trinidad Tectonic Zone (CTTZ; two AQS Tracts), with 39 ML/day (9 imgd) of combined, untapped groundwater potential. In addition, one hybrid system (Northern Valley Gravels) that rests on metamorphic aquifers and two marginal lithologies in the Central Range (Tamana limestones and Pointe-à-Pierre sands) have been delineated, with 20 ML/day (4 imgd) of combined, untapped groundwater potential.

TABLE 14. Summary of Surface Area, Recharge, and Existing Production in Megawatersheds and Aquifer Systems in Trinidad

Aquifer System/Megawatershed	Surface Area km²	Minimum Recharge (ML/day)	Maximum Recharge (ML/day)	Average Recharge (ML/day)	Existing Production (ML/day)
MEGAWATERSHEDS					
Tucker Valley	15	12	18	15	1
Maraval	82	44	53	48	23
Cascade	61	40	44	42	2
Caura	48	16	18	17	
Aripo	50	86	122	104	
Balandra	29	150	156	153	
Total	286	349	410	379	26
AQUIFER SYSTEMS					
Southern Basin					
Erin Syncline	206	42	51	46	18
Morne L'Enfer Syncline	50	8	14	11	1
Fyzabad Syncline	31	3	7	5	
Siparia-Papure Syncline	368	47	92	69	18
Moruga Syncline	118	14	23	19	
Ortoire Syncline	299	58	118	88	2
Pilote Syncline	135	22	45	33	1
Balata Syncline	32	8	12	10	
Mayaro Coastal Plain	30	5	9	7	
Total	1269	206	370	288	39

Nothern Basin

Greater Caroni Monocline	148	35	56	46	88
Couva-Claxton Bay Monocline	28	5	6	5	2
Carapichaima Monocline	48	7	16	12	24
Longdenville-Las Lomas Monocline	62	12	23	18	3
Carapo Syncline	11	2	4	3	
Mahaica Cumuto Syncline	8	2	4	3	8
Oropouche Monocline	66	36	38	37	
Guaico-Fishing Pond Syncline	85	35	39	37	
Total	454	135	186	160	124
Central Tectonic Zone					
Central Range North Flank	133	14	25	22	
Caigual Syncline	35	13	20	17	
Total	168	27	45	39	
Miscellaneous Aquifers					
Central Range Limestones	18	5	8	7	
Lower Tertiary Sandstones	25	10	15	13	
Northeast Fractured Aquifer	351	394	461	428	1
Northwest Fractured Aquifer	117	76	81	79	
Southeast Fractured Aquifer	56	22	30	26	
Southwest Fractured Aquifer	172	74	91	83	37
Total	739	581	686	634	37.30
TOTAL				1519	452

SUMMARY OF GROUNDWATER RESOURCES

Introduction

This section addresses groundwater resources, reveals the natural processes that influence aquifer formation and performance in Trinidad, and describes the genesis and taxonomy of the most significant, sustainable groundwater sources on the island. As defined and applied in this Trinidad report, an "aquifer" is a geologic environment containing naturally recharged and economically recoverable groundwater of sustainable quantity and quality suitable for practical, beneficial long-term use by municipal, industrial, or agricultural consumers.

Four types of aquifers are identified in Trinidad, and four key categories of information (discussed in the following paragraphs) are required for the accurate characterization of the potential for aquifers to meet long-term sustainable, and economically viable, withdrawals that are required by municipal, industrial, and agricultural groundwater consumers.

Aquifer Characterization

Key categories of information required for accurate characterization of aquifer potential include the following:

- Amount of recharge available to the aquifer
- Hydrogeologic (e.g., geological parameters, water storage, transmission) properties of the aquifer
- Knowledge of the geographic boundaries (e.g., perimeter) of the aquifer associated with active recharge to or functional storage of the aquifer
- Mineralogical and geochemical properties of host rocks or sediments

Aquifer Recharge

The amount of recharge available to the aquifer is locally a function of geology (e.g., lithology, stratigraphy, structure, geochemistry, geomorphology), land cover (e.g., farm lands, vegetation, urban asphalt), and climate (e.g., precipitation, insolation), which control the "hydrogeologic properties" of the aquifer. Major aquifers are often also influenced by regional tectonic forces, whose structures overprint and alter local geology, natural land cover, topography, and climate to create more favorable environments for aquifer formation, enhance aquifer boundaries, and recharge catchment.

Hydrogeologic Properties

Hydrogeologic properties that influence aquifer performance comprise a variety of structural and stratigraphic characteristics that are intrinsic to the rocks and sediments in this study. These properties exist within the larger physical context of the topographic setting, which often exerts strong control on the location of recharge and discharge regions, and the topographic relief, which often controls precipitation amounts (e.g., "orographic effect") and surface and subsurface hydraulic gradients that drive groundwater flow.

Geographic Boundaries

An aquifer is contained within geographic boundaries defined by the permeability of the host rock. Regardless of its porosity, and in order to meet the definition of an aquifer as applied in this report, the host rock must also possess sufficient permeability to allow ready access to active recharge and sufficient storage if it is to provide the requisite daily water yield to consumers through dry periods without degrading the water quality or long-term sustainable yield of the aquifer.

Mineralogical and Geochemical Properties

During aquifer evaluation, the mineralogical and geochemical properties of the host rocks are as crucial a factor to groundwater explorers as they are to other economic geologists who seek such precious commodities as oil, gold, and diamonds. The mineral and chemical makeup of the unconsolidated or consolidated aquifer host rocks exert a strong control on the amount of porosity and permeability, and the chemistry (e.g., iron, manganese, and hardness amounts) of the aquifer.

Aquifer-Host Geological Environments in Trinidad

Geological environments containing aquifers in Trinidad include Northern Range metamorphic rocks and unconsolidated sediments. Within these two broad categories, there are subclasses of aquifer types related to the heterogeneous and complex geology of Trinidad.

Northen Range Metamorphic Rocks

Northen Range metamorphic rocks in Trinidad evidences two aquifer types, structural (i.e., fracture-hosted) and carbonate, as discussed below. Key structural features vital to bedrock aquifer formation include folds and faults at the regional scale and a variety of structures at the local scale, including joints, bedding partings, foliation partings, and small-scale brittle fractures. The lithostratigraphic sequence (stratigraphy based on rock type and composition) influences the thickness and chemical character of aquifers and is a controlling factor in the formation of confining beds (e.g., aquicludes and aquitards) and confined, or artesian, aquifers.

Both structure and lithostratigraphy often interact to influence aquifer formation and performance. For example, folds or faults may control the location of carbonate rock units relative to other rocks, and relatively calcite-rich carbonate rock units have the potential to be partly dissolved by groundwater. In cases where acidic meteoric waters have flowed in contact with these units over geologic time, void space is created by dissolution; and in cases where structural features (e.g., faults and fractures) conduct meteoric waters into and through soluble bedrock units, permeability is considerably enhanced. Based on the physical nature of the pore spaces or permeable fracture networks that can be sufficiently interconnected to provide suitable, laterally extensive permeability and hydraulic conductivity, two types of consolidated bedrock aquifers are exist and are described below.

Fracture-Hosted Aquifers

The competent rocks of Trinidad include carbonates and metamorphic units, which derive secondary porosity (also commonly called secondary permeability and fracture permeability) characteristics from cleavage along foliation and planar fractures pervading the bedrock matrix. In Trinidad, the main fracture-hosted aquifers are located in Northern Range metamorphic rocks that have sedimentary protoliths. In these protoliths, metamorphism in these rocks has destroyed all primary porosity and permeability and original grain shape and size. Original protolith mineralogy has been altered to greenschist facies metamorphic minerals, and because metamorphism does not alter the elemental composition of the rocks (it only changes the arrangement of those elements and the percentages in which they occur), protolith mineralogy has a strong influence on metamorphic rock composition and the development of metamorphic cleavage. The latter may play a major role in recharge infiltration and bulk hydraulic storage in these rocks.

Carbonate Aquifers

In the Northern Range, some carbonate rock units derive tertiary porosity from hydraulically connected systems of dissolved rock pore spaces and possess significant secondary porosity. In fact, much of the dissolution of carbonate rocks will tend to focus along fractures, bedding planes, and foliation planes, hence the formation of extraordinarily productive aquifers and megawatersheds in Trinidad Northern Range carbonate schists, which have produced from single bedrock wells drilled during the production phase of this project up to 10-million ML/day (2.2 imgd) sustainable yield. This type of aquifer, (#4, discussed below) discovered and documented here for the first time anywhere, is 15 to 30 times more productive than the maximum yield expected from metamorphic rocks [Driscoll, 1986].

Unconsolidated Sediments

Unconsolidated rock environments derive their primary porosity from intergranular pore spaces. Initial geological factors controlling the quality of sedimentary aquifers are intrinsic permeability, porosity, size, shape, internal heterogeneity, and external continuity. All of these are ultimately controlled by sedimentological hydrodynamics and clast mineralogy at the time of deposition. In this study, the former is simply termed depositional environment and the latter mineralogy. Following deposition, aquifer quality is also affected by diagenetic processes, which in general act to decrease primary porosity and permeability. Mineralogy plays an important role in diagenesis and also affects water quality in the aquifer. For example, the conversion of iron oxide to iron sulfide is a diagenetic process that is responsible for the high iron concentrations in many aquifers in Trinidad. In Trinidad, the major sedimentary aquifers have not undergone significant postdepositional diagenesis.

Where faults are often an aquifer-enhancer in Trinidad's metamorphic rocks, faults cutting through unconsolidated sediments in Trinidad tend to displace the stratigraphy, thereby disrupting groundwater flow in strata-bound aquifers. This phenomenon effectively forms compartments within such aquifers, which constrains recharge and storage and reduces the safe yield.

The unconsolidated aquifers of Trinidad can be classified into two major groups. Coarse-grained alluvial gravels that have their source from the weathering of the Northern Range and fine-grained sands that have a multitude of sources.

Coarse Alluvial Aquifers

Coarse alluvial aquifers are present at the base of, and in many stream valleys within, the Northern Range. They consist of sands and gravels that are composed of lithic fragments of the Northern Range metamorphic rocks. The permeability of these aquifers varies depending on the transport distance from the source to the site of deposition. The cleaner and better sorted the sediments, then the higher the permeability. These coarse alluvial deposits vary from being colluvial (poorly sorted and dirty) to very clean alluvial gravels. Generally, the gravels increase in permeability, going from the east to the west along the Northern Range. This relationship is also true for the Northern Range valley alluvials.

Fine-Grained Sand Aquifers

Fine-grained sand aquifers are the predominate sedimentary aquifers of Trinidad. They are primarily sourced from the Orinoco River in South America, the Northern Range, and the Central Range. These sediments are very fine-to medium-grained sands that were generally deposited in a shallow water beach to deep water slope environment. In Trinidad, the permeability of these fine sands varies with the quantity of silt and clay between the pores of the sand. Generally, the deep water sediments have more clay within the matrix; thus, they tend to be less permeable. The aquifers in the Northern Basin tend to have a higher permeability than the Southern Basin sedimentary aquifers primarily because they are formed in a shallow water environment.

Northern Range Rock Aquifers: Metamorphic Fracture/ Dissolution Systems

Fracture Porosity, Permeability, and Hydraulic Conductivity   Fracture-hosted aquifers occur in a variety of competent rocks exhibiting a range of matrix porosities, from negligible (e.g., unfoliated metamorphic rocks) to very high (e.g., uncemented sandstones). In all cases of aquifer development, groundwater storage also occurs within a planar fracture network exhibiting fractal architecture, with variable orientation, length, aperture, surface roughness, spacing, and degree of connectivity to other fractures. Within any individual or adjacent rock units pervaded by fractures, a series of subparallel fractures is called a fracture set. Multiple intersecting fracture sets with variable strikes are referred to as fracture systems.

In many types of soluble bedrock, with or without primary porosity, geographically extensive physical connectivity of fracture systems provides ready pathways for near-surface, acidic meteoric water infiltration, which enhances fracture size, aquifer permeability, hydraulic continuity, and results in very high hydraulic conductivities. Groundwater flows along interconnected networks of such fracture systems can create "megawatersheds," which extend groundwater catchments far beyond local topographic constraints, resulting in sustainable groundwater sources orders of magnitude greater than expected using traditional hydrogeological models.

In the Northern Range of Trinidad, joints are commonly observed at outcrop scale and late-stage brittle deformation formed within regional fracture systems. Such fracture systems were observed by ETI geologists on satellite imagery, aerial photographs, topographic maps, and digital elevation models (DEM), where they appear as linear or curvilinear features, called "lineaments." Field mapping confirmed that many of the lineaments in Trinidad are long, deep zones of closely spaced extensional fractures and fracture systems, or contacts between rocks with contrasting erosional resistance, but with elevated secondary permeability, causing these lineaments to have importance as groundwater exploration targets, potentially forming core groundwater transmission zones of megawatersheds.

In fractured aquifers, the storage may be quite uniform across scales, but the hydraulic conductivity within a single aquifer will often vary by at least several orders of magnitude across scales because relatively larger volumes will include higher order fracture types that may not be intersected in a given smaller volume. Across the entire spectrum of igneous, metamorphic, and sedimentary rocks (but excluding cavernous carbonate rocks), fracture-related hydraulic conductivities span about nine orders of magnitude from $\sim 10^{-8}$ m/day to ~ 10 m/day [e.g., Heath, 1983]. The hydraulics of large-scale, fracture-hosted aquifers, or megawatersheds, are assumed to approximate an anisotropic, homogeneous hydraulic medium and are often modeled accordingly.

In order to be productive enough to meet the requirements of the ETI-Lennox Trinidad Groundwater Development Project, groundwater supply wells in fractured rock must intercept (1) many fractures of low aperture but high connectivity to the overall fracture system and surface recharge, (2) a single fracture (or fault) of critically large aperture and high connectivity to the overall fracture system and surface recharge, (3) several fractures of moderate aperture and high connectivity to the overall fracture system and surface recharge, or (4) any of the above conditions with high connectivity to rocks possessing primary porosity and surface recharge. All of these conditions have been identified by ETI at various locations in the Northern Range, where tectonic stress regimes have produced brittle deformation, inducing faults and fracture sets in the bedrock and creating regionally permeable environments in very high rainfall areas, resulting in the formation of fractured bedrock aquifers and several megawatersheds with prolific groundwater development potential.

Dissolution Enhancement of Porosity/Permeability   Carbonate aquifers showing tertiary porosity may also contain secondary porosity. If karst landscape features such as sinkholes, sinking streams, and caves have developed, some recharge of the carbonate aquifer will be concentrated at particular point inputs. However, more diffuse recharge also occurs between the point inputs. Groundwater flow within such karstified aquifers may be fast within interconnected conduits, but between such conduits the aquifer secondary porosity may be lower, resulting in relatively slow, diffuse flow. Carbonate aquifers characterized by only secondary porosity may have hydraulic conductivities ($\sim 10^{-4}$ m/day) approximately eight orders of magnitude lower than conductivities in karsted, mature cavernous aquifers (to $\sim 10^4$ m/day). In such aquifers, the mean hydraulic conductivity increases with measurement scale up to a spatial limit related to the amount of dissolution that has occurred. As dissolution proceeds in such aquifers, heterogeneity and the aquifer volume necessary for modeling equivalence to a homogeneous medium both increase; most carbonate aquifers are anisotropic. Even in highly cavernous, mature carbonate aquifers, most of the storage ($\sim 95\%$) occurs in the fracture systems; however, most of the flow occurs within interconnected dissolution conduits.

Therefore, bedrock aquifers with extremely high local conductivities are not by themselves good targets for sustainable, high-quality water supplies because water runs too quickly through them for long-term use, but when they form part of a much more extensive, fractal network of groundwater conduits benefiting from primary and secondary porosity and draining large rainfall catchments (i.e., megawatersheds), then strategically located and constructed wells will optimize both storage and well efficiencies, resulting in highest benefit/costratio "super wells."

Northern Range Fracture/Dissolution Aquifers

Fractures or dissolution can form aquifers that pervade several bedrock units with similar minerologies or brittle fracture responses to common stress fields, such as tectonic forces. Based on these shared structures, the rocks can be grouped together and considered as a single fracture aquifer. This concept of fracture/dissolution aquifers is important to the metamorphic rocks of the Northern Range because its different rocks and their structures can be subdivided into numerous potential fracture/dissolution aquifers. The often well-developed foliation in these rocks may also act as a pathway along which brittle fractures and joints develop. This process and its importance is discussed in the following paragraphs.

Prior to metamorphism, protolith rocks in the Northern Range were primarily sedimentary with the exception of the Sans Souci metavolcanics, where the protolith rocks were a mixture of predominantly volcanic (extrusive igneous) rocks, plus lesser plutonic (intrusive igneous) and sedimentary rocks. Sedimentary protoliths of the Northern Range include mudstone to shale, siltstone, sandstone, and limestone. Low-temperature greenschist grade metamorphism (~250 to 400°C) produced the slate, phyllite, schist, metasandstones (termed quartzites), and metalimestone (a type of low-grade marble), currently exposed at the surface of the Earth. Varieties of each of these main metamorphic rock types are found in the Northern Range of Trinidad.

Many Northern Range metamorphic rocks contain a tectonic foliation defined by the preferred alignment of minerals, especially micas, that formed during metamorphism and ductile deformation of the rock mass. This foliation is called slaty cleavage in slates and generally referred to as schistosity in the other foliated rock types. Some of the metasandstones and metalimestones are nonfoliated over thicknesses of ~10 cm to 3 m, whereas in other rock types, the foliation spacing ranges from several millimetres to penetrative or microscopic.

During postmetamorphic, brittle tectonic deformation of these rocks, it is possible for them to locally undergo extension (dilation) along the cleavage or schistosity foliation planes. This process does not involve the formation of new fractures, but rather it involves parting of the rock along these preexisting foliation planes. This foliation parting can locally provide increased porosity to the rock mass and aquifer. The tectonically induced foliation parting may occur along the walls of brittle faults and the walls of extension fractures, as these are locations at which extension may pull apart the foliation. Similar extension may also occur during erosional exhumation of foliated metamorphic rocks (the western part of the exposed Northern Range was formerly ~10 km below the present ground surface) as the rock mass tends to expand and foliation parting may occur along the upper bedrock, especially where the foliation is at low angles to the eroding land surface and the maximum amount of weight is lifted off the foliated rocks. Such unloading-induced foliation parting may increase the amount of infiltrating, recharging water that is able to penetrate the upper parts of the bedrock.

Systems of extension fractures (joints) and brittle faults have formed during relatively late (postmetamorphism) brittle deformation of the Northern Range rocks. Owing to the large area of the range affected by these structures, the tectonic stresses that produced them were likely regional in scale. Some of these brittle fractures are related to regional tectonic stresses; some also may have been caused by uplift and erosion. The most systematic and physically consistent joints appear to have developed in the metalimestones (e.g., the "bluestones" within the Maraval carbonate complex) and the most quartz-rich metasandstones (e.g., some of the quartzites within the Maracas complex). These two rock types generally support the most well-developed joints because they have very high percentages of calcite (in the metalimestones) and quartz (in the metasandstones). Calcite and quartz are minerals that have relatively high compressive strengths, or the ability to resist tectonic compressive forces, and they are therefore not likely to develop foliations and more likely to develop joints. In contrast, other minerals common in Northern Range metamorphic rocks (mostly muscovite, chlorite, paragonite, pyrophyllite, graphite, feldspars, and clays formed from the breakdown of feldspars) are easily foliated and less likely to support jointing. Also, in calcite and quartz, recrystallization and pressure solution that accompanied metamorphism cause these minerals to often form highly interlocking grain networks. As a result of these conditions, the calcite-rich metalimestones and the quartz-rich metasandstones (quartzites) are poorly foliated (with respect to the slates, phyllites, and schists) and are considered stronger (more competent) than the foliated rocks. Therefore, the metalimestones and metasandstones are more susceptible to developing fractures and joints. During postmetamorphic brittle extensional deformation, these rocks were able to store considerable elastic strain energy prior to fracturing. Storing this energy meant that once these rocks started fracturing (forming joints), then the stored energy, once released, acted as a driving force that developed relatively long, persistent joints.

Joints in the nonmetalimestones and nonmetasandstones are generally less numerous and continuous because these rocks fracture at lower levels of stored strain energy. These other rocks store less strain energy because they consist of less calcite and quartz than do metalimestones and metasandstones, and because well-developed foliation makes them weaker to stress applied oblique to the foliation plane.

In Trinidad bedrock (in the Northern Range), joints are usually observed at outcrop scale. In addition to developing joints, late brittle deformation has formed long arrays of fractures that are often regional in scale. These zones of fractures are called lineaments and are conspicuous features on the SLAR satellite image and the large-scale DEM of the Northern Range metamorphic rocks (discussed previously in Chapter 4). Most lineaments are continuous and mark the traces of steep to vertically dipping fractures along which no offsets can be detected. They are often long, deep zones of closely spaced extensional fractures, or they are otherwise defined by contacts between rocks with contrasting erosional resistance. Most of the lineaments do not appear to be faults. There is a correlation between many joint (fracture) patterns and the larger-scale lineaments in the region where they occur.

The fractured zones are zones of elevated secondary permeability, and these lineaments have potential economic importance in two main types of exploration: (1) as channels along which ore-bearing fluids can flow, and exploration tracing along the lineaments may lead to locations of ore deposits; and (2) as channels for groundwater to flow, thus possibly forming the core zones for some types of megawatersheds. Groundwater supply wells sited along lineaments may provide especially productive and sustainable withdrawals; of course, not all geographic areas that need additional water will be in the vicinity of a lineament.

Lineaments are very likely a manifestation of large-scale mechanical segmentation of the crust. The exact mechanism of formation is not known, but some lineaments may form from dilation (expansion) rather than distortion (shape change) of the crust. One mode of dilation may involve expansion of formerly deeply buried rocks that have been uplifted to shallow levels. Earth tides may cause stress fatigue and fracturing of rocks that may be focused along lineaments. Deep-seated, regional crustal motions may form some lineaments.

The Megawatershed Paradigm   The "Megawatershed Paradigm" is a conceptual model describing natural complex systems of water catchments and drainage genetically linked to tectonism and consisting of three-dimensional surface and subsurface "zones" often evidencing active interbasin groundwater transmission in consolidated fractured bedrock and compartmentalization in faulted, deep sediments. This model supercedes the traditional watershed model and has profound implications in river basin models, which most often describe topographically controlled, functionally two-dimensional drainages and depict deep groundwater resources as static, poorly recharged artifacts of surface flow, confined to discreet bedrock or alluvial units.

The Megawatershed Paradigm describes deep groundwater systems' water catchments, recharge, and transmission and storage boundaries beyond the limits of traditional watersheds. Structural geometries incompassing many watersheds are defined by tectonically induced, large-scale, highly variable fracture sustenance. In defining rainfall catchments, groundwater flow mechanisms in consolidated and unconsolidated rocks may frequently transcend synthetic watershed and basin boundaries in consolidated rocks or, conversely, may prevent lateral groundwater flow by previously unsuspected stratigraphic offsets also associated with tectonism.

The use of the Megawatershed Paradigm as a groundwater exploration model, as is the case in this Trinidad Project, greatly improves the understanding of groundwater environments. From this understanding, accurate and comprehensive assessments of "perennial yield," also called "safe yield" from local, country-wide, or regional water resources can be deduced. The model actually provides a template for measurement of all facets of the hydrologic cycle in reality-based natural hydrogeological catchments, including rainfall, evapotranspiration, surface and subsurface inflow and outflow interactions with surface water, shallow aquifers, deep fractured bedrock, and deep alluvial aquifers.

Northern Range Fracture/Dissolution Aquifer Systems
Megawatersheds and Fractured Rock Aquifers

In the predominantly metamorphic rocks of the Trinidad Northern Range, most of the primary porosity has been obliterated and secondary (fracture) porosity has been developed, which is further enhanced by tertiary (dissolution) porosity within the meta-carbonate rocks. Much of the slow dissolution of the metacarbonate rocks will tend to focus along fractures (including joints, outcrop-scale faults, and regional-scale faults) and the tectonic foliation. Under the weathering conditions of Trinidad, the mineral quartz will also dissolve, but the rate of quartz dissolution is extremely slow as compared with calcite dissolution, and quartz dissolution porosity within bedrock is generally expected to be negligible.

Based on the degree of susceptibility of rock type to (1) fracturing, (2) calcite dissolution, and (3) creation of clay from the weathering and decomposition of primary minerals such as feldspar, the Northern Range aquifers are classified as having excellent, good, fair, or marginal aquifer potential. Table 15 shows the distribution and ranking of various Northern Range lithologies.

In the western half of the Northern Range, there are large alluvial valleys where the alluvium is predominantly gravel or clean sand and some of the most prolific alluvial wells within the country are located (Tucker Valley and Diego Martin). Toward the east, the valleys narrow with steeper valley walls, resulting in the deposition of colluvium and poorly sorted sand and gravel. Consequently, the productivity of water wells is much lower. However, this lower permeability material acts as a good storage media available to underlying fractured bedrock aquifers for recharge.

The likelihood of achieving sufficiently large yields from groundwater supply wells in Northern Range bedrock is increased by situating the wells along zones of secondary and tertiary permeability that is the result of combining favorable rock types with intense fracturing. In the Northern Range, the most favorable rocks are those that contain carbonate and the least favorable are those that have an abundance of phyllite and mica accompanied by a lack of quartz. Areas with an abundance of fracturing are usually topographically low areas that can be identified by the analysis of remotely sensed imagery such as aerial photographs, satellite spectral imagery, side-looking airborne radar, and satellite radar. Other methods that are used include topographic analysis using digital elevation and digital raster graph data. For rock types with no susceptibility to dissolution, productive wells in bedrock must tap a large number of transmissive fractures that are part of an interconnected fracture system within the surrounding rock mass and aquifer. In some cases, there are very long, wide-aperture (highly gapped) extensional fractures or faults that have the capability, under sustained pumping stresses, of drawing groundwater from relatively great distances that reach beyond the surface watershed within which the well is located. Even when these features occur within poor quality bedrock types, very high yield wells can be developed. ETI drilled a very high yield well (several hundred gallons per minute) within the graphitic phyllites in the Valencia area.

TABLE 15. Ranking of Northern Range Lithologies

Formation Number	Formation Name	Lithologic Ranking	Well Yield (igpm)	Well Yield (ML/day)
30	Galera Grit	Good	350–700	0.11–0.22
31	Sans Souci Metabasalt	Good	350–700	0.11–0.22
32	Toco Melange	Good	350–700	0.11–0.22
33	Unnamed Slate and Metasandstone	Marginal	20–100	0.006–0.03
33ls	Metalimestone Layers	Good	350–700	0.11–0.22
33s	Metasandstone Layers	Marginal	20–100	0.006–0.03
34	Unnamed Calcareous Slate, Phyllitic Slate & Metalimestone	Excellent	>700	>0.22
34ls	Metalimestone Layers	Excellent	>700	>0.22
35	Unnamed Phyllite, Phyillitic Schist and Metasandstone	Marginal	20–100	0.006–0.03
35ls	Metalimestone Layers	Good	350–700	0.11–0.22
35s	Metasandstone Layer	Fair	100–350	0.03–0.11
36	Unnamed Quartz and Quartz-Mica Schist	Fair	100–350	0.03–0.11
36ls	Metalimestone Layers	Good	350–700	0.11–0.22
37	Maraval Carbonate Schist and Calcareous Mica Schist	Excellent	>700	>0.22
37ls	Metalimestone Layers	Fair	100–350	0.03–0.11
38	Maracas schist	Fair	100–350	0.03–0.11
38q	Metasandstone Layers	Fair	100–350	0.03–0.11
39	Chancellor Schist	Good	350–700	0.11–0.22
39ls	Metalimestone Layers	Good	350–700	0.11–0.22
40	Laventille Metalimestone	Good	350–700	0.11–0.22
41	Lopinot Phyllite	Marginal	20–100	0.006–0.03
41ls	Metalimestone Layers	Marginal	20–100	0.006–0.03
42	Morvant Metasandstone	Marginal	20–100	0.006–0.03
43	Cataclasite and Protocataclasite	Good	350–700	0.11–0.22

The hydrogeologic characteristics of the Northern Range lithologies form the basis for their aquifer rankings outlined below. These rankings comprise five categories: excellent, good, fair, poor, and marginal bedrock types. The megawatersheds have been defined within those bedrock types that are ranked as excellent. The hydrogeologic map of Trinidad that accompanies this report shows the distribution of these various units in the Northern Range. Table 15 summarizes the lithological ranking and lists the anticipated yields of precisely located wells within each formation.

Excellent Lithologies

Maraval Carbonate Schist and Calcareous Mica Schist (Unit 37) Karst features are locally well developed within the metalimestones of the Maraval complex (visible on the SLAR and DEM images). Many caves in the central Northern Range (e.g., Aripo, Cumaca, Oropouche) are present within these low-metamorphic grade Maraval marbles and their counterpart lithologies to the east (the latter are within the vicinity of, and east of, the Hollis Reservoir, where on our ETI Northern Range geologic map, we refer to them as the calcite-rich units within the "unnamed calcareous slate, phyllitic slate and metalimestone"). The metalimestones may generally be excellent lithologies for the development of aquifers because they tend to support widely well-developed fracture permeability; locally, the foliation planes and the joints have been dissolved to develop extensive supplemental dissolution porosity and permeability. The carbonate schists are generally good lithologies showing limited jointing and less, but significant, dissolution porosity.

Unnamed Calcareous Slate, Phyllitic Slate, and Metalimestone (Unit 34) This unit is very similar to the Maraval Carbonate Schist and is probably an equivalent unit but at a lower metamorphic grade. Refer to the description above for the Maraval Formation. Portions of these rocks may produce excellent aquifers because of secondary porosity caused by dissolution of calcite within metacarbonate layers and calcareous metamudstones, which is evident by the existence of the numerous caves within the Northern Range. This unit extends to the east all the way to Balandra Bay.

Good Lithologies

Laventille Metalimestone (Unit 40) The metalimestones are of lower metamorphic grade than those observed elsewhere in the Northern

Range, and they are more similar to limestone than low-grade marble (metalimestone). Weak to moderate karst development occurs locally in these rock units. These metalimestones are highly fractured, but many fractures are closed via vein-filling by secondary minerals. Overall, the rocks are ranked as good lithologies for the development of bedrock aquifers. The metamudstone interlayers are ranked as fair, and because they are interlayered with the metalimestones, they may limit hydraulic communication between the metalimestones. The metamudstones generally show a spaced cleavage oriented at an angle to bedding, which may foster slow groundwater movement along the foliation and across the beds.

San Souci Metabasalt (Unit 31)   Because of the high fracture, fault, and shear-zone density in this unit [e.g., Wadge and Macdonald, 1968], this predominant metabasaltic rock will very likely result in the formation of fair-to-good aquifers. Interlayers of metamudstone may serve as aquacludes, but many of these layers are laterally discontinuous, thereby allowing some movement of water across these layers. Overall, the San Souci complex is likely to have high groundwater storage and three-dimensional hydraulic communication, as thus is ranked as a good formation for the development of groundwater.

Chancellor Schist (Unit 39)   The quartz-mica schists and carbonate schists will generally host fair-to-good aquifers because where well jointed, the metalimestones have both potential fracture and dissolution porosity and permeability that would be characteristic of an excellent aquifer. However, the metalimestones are a lesser component of the overall Chancellor complex, and the rather isolated outcrops of this lithology apparently do not support karst features. The metaquartzites may locally support well-developed fracture permeability and form good aquifers; however, these units are thin and of limited surface extent. The quartz mica schists will display limited jointing and low levels of porosity and permeability. However, video camera surveys and very high production rates of an ETI production well have shown that the carbonate schists have enhanced permeability because of dissolution of the carbonate along mineral boundaries and the axes of carbonate replacement fold structures.

Fair Lithologies

Maracas Schist (Unit 38)   The quartz-mica schists generally host marginal-to-poor aquifers because of low porosity and permeability. It is likely that pore spaces in the protoliths have been largely filled by the metamorphic recrystallization of quartz and mica. The schistose foliation causes these rocks to be resistant to water flow normal to the foliation plane. However, slow water penetration through these rocks may occur along quartz-rich panels parallel to, and located between, the mica-rich foliation planes, allowing foliation-parallel groundwater movement. The metasandstones (quartzites) will generally host fair aquifers because these rocks contain more abundant systematic joints than do the schists (well-foliated rocks). The joints in these units are generally at high angles to the foliation. The greater abundance of joints within the metasandstones (quartzites), and the lower abundance of joints found within some of the quartz-mica schists, have the potential to form fracture permeability networks allowing groundwater to flow across (or perpendicular to) the foliation. Overall, this formation is considered to have fair aquifer potential.

Unnamed Quartz and Quartz-Mica Schist (Unit 36) Local metasandstones (quartzites) occur within this complex and may show fairly well-developed joint systems at high angles to the foliation that render them fair aquifers. However, these metasandstones (quartzites) are typically thin and will not have large groundwater storage. The quartz-mica schists are marginal-to-fair aquifers that generally have limited joint development and low fracture porosity and permeability. This formation is very similar to the Maracas schist above and is thus ranked as a fair lithology.

Galera Grit (Unit 30) Galera unit rocks are ranked as a lithology that forms fair aquifers. Quartz cementation and subsequent recrystallization have closed many of the pore spaces and lowered the intergranular permeability between individual quartz grains. However, locally well-developed joint systems may form moderately productive fracture permeability networks.

Marginal Lithologies

Morvant Metasandstone (Unit 42) These slates and metasandstones (quartzites) have small surface extent and locally are cut by the Arima Fault. Hydraulic continuity in the Morvant units may be disrupted by large amounts of displacement associated with these faults. The slates are likely to be marginal aquifers that retard hydraulic communication between the metaquartzites. Where the metasandstones (quartzites) show well-developed joints they may form fair aquifers, but the very limited extent of the Morvant units suggests they are low-ranking candidates for groundwater exploration.

Unnamed Slate and Metasandstone (Unit 33)   This unit is very similar to the Morvant metaquartzite phyllite and Slate except it is at a much lower metamorphic grade. We consider this a marginal lithology for the development of a bedrock aquifer because of the clay-rich composition and its tendency to take up tectonic stress by movement along mineral boundaries that become smeared into microfault gouge because of the softness of the material instead of creating open fractures across the foliation.

Lopinot Phyllite and Slate (Unit 41)   The Lopinot complex will generally form a marginal aquifer. Permeability across and along foliation will be very low, owing to the clay- and mica-rich composition and very fine grain size of these rocks. However, where this unit is located in an intensive tectonic zone, it can be favorable. ETI test drilled on the intersection of the Arima fault and a northwest trending lineament within this formation, in the Valencia area, developing over 0.5 imgd (VT-2). However, ETI also drilled in areas where this structural regime is not present, resulting in the development of very low-yielding wells (less than 10 igpm; GT-1, GT-2).

Unnamed Phyllite, Phyllitic Schist, and Metasandstone (Unit 35)
These undivided rocks are assessed as marginal lithologies for the development of bedrock aquifers. Although less metamorphosed than the schistose rocks to the west, these rocks contain a spaced mica-rich foliation, slaty cleavage, and original clay-rich beds that will serve to impede groundwater movement.

Sedimentary Basin Fill Systems

Aquifer Quality Sediments (AQS)   The concept of aquifer quality sediments (AQS) is central in exploring for and developing groundwater resources using sedimentological, sequence-stratigraphic, and basin analysis methods. AQS, which include sand and gravel, also must have sufficient intrinsic permeability to provide the transmissivity necessary for the installation of production wells of sufficient yield to make them economical. From studies of process sedimentology and facies analysis in modern and ancient environments, it becomes clear that high intrinsic permeability sediments (AQS) are relatively uncommon, although they may be locally abundant. Bodies of AQS are deposited in a limited number of environments, each having predictable lateral extent and distinct depositional geometries. Understanding how AQS bodies are formed and preserved is one of the main elements of successful groundwater exploration. Therefore, the AQS concept is a central theme in the following discussions of sedimentary aquifer systems.

Intrinsic Permeability, Hydraulic Conductivity, and Primary Porosity   The most important factor for determining whether a particular gravel or sand is capable of performing as an aquifer is the intrinsic permeability of the sediment. Permeability is a measure of the ease with which flow can occur among the pores and through the pore throats of a sediment. By considering fluid viscosity and density and by incorporating the effect of gravity, an expression for hydraulic conductivity (K) may be derived as follows:

$$K = \text{permeability} \times (rg/m),$$

where r is fluid density (kg/m^3), g is acceleration caused by gravity (m/s^2), and m is fluid viscosity (dyn.s/cm^2). For a single phase system, hydraulic conductivity (K) may be expressed in units of m/day, cm/sec, or gal/day/ft^2. As gravity is universally constant and for any given temperature, such as mean annual surface groundwater temperature, r and m are also constant, it should be clear that intrinsic permeability is the controlling variable for hydraulic conductivity (K).

Primary permeability through pore throats in sands and gravels is related to sediment grain size, sorting, shape, and angularity [Beard and Weyl, 1973]. Primary permeability is reduced through constriction of pore throats during cementation of grains, or less commonly but particularly in carbonate rocks, increased through grain dissolution. Grain size determines the size of pore throats and the surface area-to-volume ratio, which determines drag. Sorting, shape, and angularity all affect packing (e.g., the way grains fit together). If grains are able to rotate and realign, then compaction increases packing efficiency and decreases permeability.

For unconsolidated sediments without diagenetic cements, grain size and grain sorting are the two most important controls on permeability. Coarser-grained sediments have higher permeability than finer grained sediments, and better sorted sediments have higher permeability than poorly sorted sediments. In general, clean sands and gravels that lack significant clay and silt have sufficient permeability to be of aquifer quality (AQS) under conditions of adequate recharge. Even small amounts of diagenetic clay distributed in pore throats have the effect of drastically reducing permeability. Silty and heavy clay sands, as commonly occur in tropical deltas, also have permeabilities far lower than even poorly sorted clean sands.

Sandstone aquifers typically have permeabilities one to two orders of magnitude higher than those of most economic oil reservoirs. Although oil is often more viscous than water, two factors account for this apparent paradox. The first is that municipal water wells must deliver one to two orders of magnitude more water per day than an economic oil well produces oil (4000–80,000 gallons of oil, as compared with 100,000–1,000,000 gallons of water per day). Secondly, many oil and gas reservoirs are deeply buried, with reservoir fluid pressures (under normal hydrostatic gradient) far in excess of those in most near-surface aquifers. This high pressure has the effect of increasing the delivery rate for any specific permeability. Therefore, although intrinsic permeability in oil and gas reservoirs is typically recorded in millidarcies, aquifer intrinsic permeabilities are more likely to be several darcies. One darcy is equal to 0.04 M/day.

In terms of hydraulic conductivity (K), a low-grade aquifer sand might have a hydraulic conductivity (K) value of 2–4 m/day (50–100 gallons/day/ft^2). For fresh water at 25°C, this is approximately equivalent to intrinsic permeabilities of 2–4 darcies. Thus, an intrinsic permeability of one darcy (equivalent K = 23.3 gal/day/ft^2 and 0.04 M/day at 25°C) may be considered a very conservative minimum permeability cutoff for AQS. Among clean sands, such lower end intrinsic permeabilities are found in very poorly sorted medium sands to moderately sorted very fine sands.

Silty sands have hydraulic conductivity (K) values ranging from 0.04 to 20 m/day (1–500 gal/day/ft^2), whereas clean sands have hydraulic conductivity (K) values ranging from 0.4 to 200 m/day (10–5000 gal/day/ft^2), depending primarily on grain size and grain sorting [Heath, 1983]. Clean gravels can have even greater hydraulic conductivity (K) values of up to 4000 m/day (100,000 gal/day/ft^2) [Heath, 1983]. The important conclusion of this discussion is that to delineate a municipal aquifer, relatively clean sands or gravels that lack disseminated depositional or diagenetic clay are needed. The sedimentology of such sands and gravels is discussed below.

Porosity, the percentage volume of void space within a sediment or rock, is an important and necessary condition for aquifer development. Porosity determines the storage potential or specific yield for water in the aquifer. For aquifers having only periodic recharge, porosity is a key concept for aquifer management. Unlike permeability, porosity is not a limiting factor for aquifer delineation. This is because porosity and permeability are positively correlated, and whenever there is sufficient permeability for AQS development, there is usually also sufficient porosity to receive any available recharge.

Considerable work has been done in the oil industry on measuring the porosity of different sands and packing arrangements [for review and seminal data, see Beard and Weyl, 1973]. The empirical maximum primary porosity for clean, well-sorted, uncompacted sands appears to be about 43% [Beard and Weyl, 1973]. The maximum porosity attainable using regularly packed equal-sized spheres is 47.64% and occurs with cubic packing [Graton and Fraser, 1935]; minimum porosity of 26% is attained with rhombohedral packing. Theoretically, grain size does not affect porosity for spherical objects. In practice, median grain size is sedimentologically correlated with other factors such as sorting, angularity, and shape, which all affect packing and thus porosity. A large data set of measured porosity, permeability and mean grain size, acquired from fine-grained sediments of the Columbus Basin by bp of Trinidad and Tobago, is discussed in Chapter 2. In general, porosity decreases with increasing sphericity, roundness and compaction, and increases with increasing sorting. As expected, primary porosity decreases with intergrain cementation.

Sedimentology and Depositional Environments

From the above discussion of permeability and hydraulic conductivity, it emerges that AQS bodies can be found wherever clean sands or gravels are deposited in sufficient quantity and thickness, and with connectivity to recharge. Depositional silt and clay, and diagenetic clay, drastically reduce K values and limit aquifer potential.

The deposition of clean sands and gravels of appreciable thickness occurs in only a limited number of environments. All such depositional environments are characterized by sufficient water or wind currents to winnow and otherwise remove fine silt and clay particles from the sediment load. Some depositional systems, for example, glacial, may be removed from consideration for a tropical environment such as Trinidad. Eolian desert deposits have also never been identified in Trinidad although beach and chenier dunes might be expected to occur in places.

Clean, high-permeability sands (2–5 darcies) may be deposited in rivers, estuaries, and siliciclastic shelves by fluvial, tidal, wave, and longshore currents, and to a lesser extent, in deep-water submarine fans by turbidity currents. Tropical deltaic sands tend to be slightly to highly argillaceous. The deltaic sands of southern Trinidad, for example, are uniformly fine grained and usually contain some silt and clay that restricts permeability, resulting in limited, moderate-yield aquifers.

Medium and coarse gravels are deposited in rivers, alluvial fans, beaches, and some submarine fan facies. Debris flow rubbles deposited in steep Northern Range valleys are matrix supported and have very high clay content in the matrix, resulting in low permeability. Such material is termed colluvium and is only suitable for limited aquifer development. Streambed gravels deposited during waning flood discharge are poorly sorted but relatively clean, and therefore, they can be developed as aquifers. Marine gravels and coarse turbidites onshore Trinidad are rare. Examples include the type Cunapo gravels of the Central Range and some facies of the Herrera sands of the Southern Basin.

In summary, most of Trinidad is composed of silt and clay derived from South America. In the south, lesser amounts of fine and very fine sand are variably present but most of this is admixed with some silt- and clay-sized material that reduces intrinsic permeability. The situation is better in the Northern Basin, where estuarine sands deposited under tidal and fluvial influence are coarser and cleaner than in the south, although they are still fine. Additionally, clean, lithic, shallow marine sands are present in the northeastern Northern Basin, and aquifer quality alluvial gravels occur along the metamorphic range front in the west.

Aquifer Quality Sediments (AQS) Tracts Concept

In this study, aquifer quality sediments that are genetically related in similar systems tracts of adjacent parasequences or sequences, and therefore have similar aquifer properties, are termed AQS tracts. This terminology provides a useful concept for discussing groups of physically and genetically similar sediments that may not be in direct hydraulic connection, and thus, they do not comprise the same aquifer or aquifer system. The Sum Sum (and Durham) sands of the Talparo Formation are an excellent example of an AQS tract. As discussed later in more detail, this regional interval of sand was not deposited as a single hydraulically connected sand sheet as previously thought. Nor do the Sum Sum sands comprise only one sequence or systems tract; in many places, there is a sand-on-sand sequence boundary within an individual AQS body, representing lowstand and highstand deposits of two adjacent sequences. In other places, highstand sands of superjacent sequences are both part of the Sum Sum AQS tract.

An AQS tract, therefore, is neither a sequence stratigraphic unit nor, strictly speaking, is it a lithostratigraphic unit. AQS tracts cannot be mapped without incorporating subjective decisions regarding continuity and inclusion, and such maps would therefore have limited value. The principal utility of the AQS tract concept is for discussing the petrophysical and aquifer properties of groups of AQS bodies deposited within a limited time interval in the same basin, under similar tectonic, climatic, and sediment provenance conditions.

Later in this chapter, all AQS tracts of Trinidad are introduced and discussed in terms of their sequence stratigraphy, tectonic setting at the time of deposition, sedimentary depositional environments, and resulting aquifer properties. Subsequent postdepositional burial, uplift, and deformation can drastically affect the lateral continuity and recharge potential of different areas of the original AQS tract. Such postdepositional structures are therefore critical in the delineation of aquifers and aquifer systems.

Basin Analysis and the Aquifer Systems Concept

The tectonic and structural geology of Trinidad is very complex, with most of the deformation affecting present aquifer distribution occurring during the Pleistocene. General postdepositional structural deformation provides the primary control on the present-day subsurface attitude and continuity of AQS bodies, and therefore upon regional groundwater flow gradients. Syn-depositional tectonics, such as normal faulting along wrench systems in the Northern Basin, tends to result in locally thickened fairways of aquifer quality sediment on the downthrown side of the fault.

Structural considerations, therefore, provide the first-order basis for mapping and dividing groundwater resources. The primary focus of structural discussion in sedimentary geology is the basin. A basin is a tectonically formed low area on the Earth's crust that is available to receive and accumulate sediments. Understanding how basins are formed, filled, deformed, and reformed is the essential business of sedimentary geologists and is termed basin analysis [Miall, 1984]. Sequence stratigraphy is a primary tool of basin analysis, but this integrative activity incorporates geochemistry, paleontology, petrography, and other disciplines.

Plate tectonics may be viewed as the science of basin formation and deformation, and sequence stratigraphy as the science of basin fill. From this perspective, it becomes clear that basin analysis provides a practical framework for the restoration and understanding of Earth history at the regional scale. The general methodology of basin analysis is treated in detail by Miall [1984; 1990] and involves placing structural and stratigraphic observations from one or several adjacent basins into a tectonostratigraphic model that is plausible in terms of regional plate tectonic evolution.

Prior to this study, the application of basin analysis techniques as an explicit *modus operandi* for groundwater resource assessment in Trinidad had not been pursued. Because of this, previous models did not adequately address the structural and architectural complexity of AQS distribution. During this study, however, the present-day structural compartments of onshore Trinidad basins were mapped and interpreted in terms of their structural evolution. Such interpretations rely heavily on the sequence stratigraphic interpretation of AQS tracts and nonaquifer facies within the compartments. For example, much of the Pleistocene deformation, as discussed in Chapter 4, disrupts even the youngest Talparo and Erin sequences and clearly post dates their deposition. The same tectonic events that disrupt AQS systems are also active during the depositional process. An example of this phenomenon is shown on NBC Line 3 (SP 1350–1400, Plate 139), where growth on the northern downthrown side of a fault indicates that the fault was active during deposition of sequences Tlp3–Tlp5 and that transport was toward the north.

After regional faulting and folding, AQS bodies that were originally deposited in relatively continuous tracts will become dislocated, exhibiting separate subsurface attitude and surface recharge potential. Major structural features, such as synclines and fault-bound monoclines, will dictate the gradient of groundwater flow within their associated AQS bodies. Such AQS bodies may, for stratigraphic reasons, have separate recharge areas and little or no subsurface hydraulic connectivity.

In this study, we use the term aquifer system for such subregional groundwater systems that are delineated by major structural features and comprise several aquifers of similar stratigraphic architecture and access to recharge. Accurate characterization and aquifer modeling of such aquifer systems requires detailed sequence mapping, using appropriate tools for high-resolution imaging of AQS bodies and genetic sequence stratigraphic correlation techniques.

The delineation of aquifer systems as groundwater resources involves the integration of detailed surface and subsurface geological mapping with a hydrological analysis of recharge potential. The assessment of recharge potential for such systems is based on computative modeling of rainfall, evapotranspiration, and runoff as discussed in

Chapter 3 and below. The original structural and stratigraphic synthesis used to delineate the aquifer systems defined in this study is summarized below.

At the outset of the study, a structural and basin history analysis of the geologic evolution of Trinidad was performed, starting with a review of existing published and unpublished studies [e.g., Kugler, 1953; Suter, 1960; Ablewhite and Higgins, 1968; Barr and Saunders, 1968; Bower, 1968; Saunders and Kennedy, 1968; Pindell and Barrett, 1990; Payne, 1991; Erlich et al., 1993; Babb, 1997; and many others]. The geologic map of Trinidad, compiled by Kugler [1961; 1:50,000 scale; see Chapter 2], served as the project geological base map. The digitized map was modified to incorporate new observations and interpretations from the disparate remote sensing and subsurface data sets.

The main AQS tracts of the Northern and Southern Basins of Trinidad were broadly known from previous work [e.g., Rohr, 1949; 1955; de Verteuil, 1968]. However, sedimentologically, these units needed to be placed within a sequence stratigraphic framework at the outcrop scale, which could then be extended to the subsurface through electric logs and seismic correlations. Considerable facies analysis of outcrops throughout the island was undertaken to delineate the depositional environments and genetic context of the main AQS tracts. The resulting observations and facies models served as "calibration" for the log facies analysis of electric logs in the subsurface.

In general, the detail of investigation for any particular area was limited by the amount and quality of the data available from that area. The methods employed in the interpretation of the seismic data from the Northern and Southern Basins were decided on after a careful examination as discussed in Chapter 2. The data quality in the Southern Basin proved to be problematic. In the western Northern Basin, the quality of the NBC seismic data was excellent and supporting electric log data were readily available. Detailed structure maps on the bases of four Northern Basin sequences were generated from the NBC data (Chapters 2 and 4). However, in the eastern part of the Northern Basin, little existing data were available. Therefore, the ETI-Lennox team conducted extensive geophysical exploration activities using CSAMT techniques that differentiate in the subsurface highly resistive (sands) from conductive (clays) materials.

From the structure maps, fault systems and fold trends were identified and interpreted through geologic time, resulting in a refinement of existing tectonic models for the basin and a significant reordering in the timing of some key events, such as the uplift and emplacement of the Central Range (Chapter 4). Finally, faults that extended to surface on the seismic lines were integrated with the surface geological map, resulting in the naming at surface of several previously undelineated fault systems. Such deeply rooted fault systems (e.g., the Couva and Caparo) form the boundaries of some aquifer systems in the basin.

A series of cross sections for each sequence in each aquifer system was constructed. Initial cross sections were made to include wells closest to the seismic lines, which permitted firm well-to-seismic tie-ins, thus ensuring the validity of the genetic correlations across faults and depositional systems tracts. Infill sections were then created in areas without seismic control and correlated in terms of systems tracts, flooding surfaces, and parasequences. This methodology formed the main basis for delineating aquifer systems, within which individual aquifers can be mapped and targeted.

Thickness and Vertical Continuity

Within individual wells, AQS tracts often consist of a number of stacked, producible AQS bodies (aquifers), vertically separated by less permeable clay-rich facies. Such aquifer intervals in vertical boreholes are distinguished, below and above, from nonaquifer-quality strata. The total thickness of the sands, gravels, and interbedded clays, from the highest producible interval to the lowest producible interval, is reported as the aquifer thickness for each well.

Within this interval, resistive sands and gravels, as interpreted by the geologist from a study of the SP, Gamma and Resistivity logs, are summed to calculate the net sand thickness. The net sand thickness is expressed as a percentage of the total aquifer thickness in the net-to-gross ratio.

Lateral Continuity

The lateral continuity of AQS bodies is invariably limited. This is because of the extraordinary conditions, in terms of energy and winnowing/sorting capacity, under which these sediments are deposited. Each of the environments and bedforms discussed above, for example, fluvial bars, beaches, and tidal dunes, have a characteristic geometry, thickness, and orientation with respect to depositional flow. Depending on the environment, rate of sediment supply, basin subsidence, and eustatic base level flux, individual bedforms and larger sand bodies may coalesce through time to form a hydraulically connected aquifer of considerable lateral extent. Although this coalescing of sand bodies is broadly true, especially at a scale of less than 10 km on passively sub-siding margins, research on reservoir distribution analysis from the petroleum industry shows that at a larger scale, this seldom holds true. New technology showing flattened 10–30 m (30–100 ft) thick "syndepo-sitional" horizons derived from three-dimensional seismic volumes, bears out visually what has been long established among subsurface geologists mapping with e-logs, namely, that fluvial, deltaic, and tidal sands have a notoriously patchy distribution. In areas with significant geologic structure, such as the synclines and monoclines of the North-ern and Southern basins, continuity of sands cannot be assumed down structural dip.

Outcropping sands, such as the Sum Sum, should not be modeled and mapped as a continuous tabular body plunging into the subsurface down structural dip. The way to accurately map such sands is by using the detailed genetic sequence stratigraphic correlation and subsurface mapping techniques developed by the petroleum industry [see Van Wagoner et al., 1990; Tearpock and Bischke, 1991]. These techniques include making detailed structural maps from seismic data on multiple horizons and integrating these with sand thickness maps to develop a three-dimensional understanding of AQS distribution in the aquifer system. To our knowledge, this is the first study of Trinidad aquifers that attempts such a procedure, applying it to some aquifers of the North-ern Basin. Although this exploration technique is possible for aquifer systems of the Southern Basin, such detailed mapping is beyond the scope of this study, which is primarily a hydrogeologic assessment of groundwater potential on the island. However, this does not mean that simple layer-cake models for estimating aquifer distribution and volume can be applied, as has been accepted practice in the past. Informed estimates of lateral sand continuity and thickness need to be made based on an understanding of the sedimentary environments and the construction of stratigraphic cross sections incorporating strategic wells, correlated using genetic sequence stratigraphic principles [Van Wagner et al., 1991].

Mineralogy and Water Quality

Sedimentary aquifers in Trinidad are composed of sands and gravels. The dominant source of aquifer sands is from the South American continent via the Proto-Orinoco and other such systems to the west. Ultimately, these sands have a Guyana Shield provenance. Most aquifer gravels in Trinidad are derived locally from the metamorphic Northern Range.

South American-derived sands, such as occur in the Morne L'Enfer, Forest, Moruga, and Talparo aquifers, are all mineralogically dominated by fine and very fine quartz. Some sands, such as those of the Morne L'Enfer, have a significant component of altered feldspathic grains. Micas and heavy minerals, including various iron sulphides, occur as accessory minerals.

Closer to the Northern Range, sands such as those at Sangre Grande, have a significant lithic component of metamorphic rock fragments, primarily graphitic phyllite. Metamorphic vein quartz is also an important component of these and other sands close to the Northern Range.

Bioclastic sands, sands containing diagenetic minerals such as glauconite, and mixed siliciclastic (quartz-rich)/bioclastic sands are the other main sand types that occur in Trinidad. None of these are significant aquifers. For example, the Montserrat Glauconitic sand is a mixed siliciclastic-bioclastic-glauconitic sediment with considerable clay content as well.

Glauconitic sands are likely to produce water quality problems in aquifers because of their high iron content. The quartz sands of Trinidad, however, are chemically stable and do not adversely affect water quality. High iron in some Trinidad aquifers may be partly caused by oxidation of detrital iron sulphides in aquifer sands, but it is mainly from iron influx during slow recharge from adjacent clays.

Aquifer gravels derived from the Northern Range are variably composed of vein quartz pebbles and cobbles and phyllitic rubble. Metamorphic rock fragments form a major component of sands associated with such gravels in stream bars and gravel fans.

Sedimentary Diagenesis

The main diagenetic process that affects sedimentary quartz sand aquifers is cementation and associated lithification. Cementation reduces the volume of pores and decreases the diameter of pore throats, with attendant reduction of porosity and permeability. Lithification has the potential advantage of increasing the rock's susceptibility to brittle fractures. From a hydrogeologic perspective, however, the attendant increase in fracture porosity and permeability seldom equals the associated loss of primary permeability and porosity. In addition, the tectonic environments needed to essentially "shatter-fracture" such sandstones are rare, and they have associated drawbacks of their own. For this reason, most truly fracture-hosted aquifers occur in regionally deformed igneous and metamorphic rocks.

Most post Oligocene sands in Trinidad were deposited rapidly in deltaic and turbidite environments. Subsidence caused by tectonism is high in Trinidad, and there is a plentiful supply of sediment from the South American continent. This combination results in geothermally cold basins with thick undercompacted sediment piles and very little diagenetic cementation. As a result, cementation and other sedimentary diagenetic processes usually play an insignificant role in Trinidad hydrogeology, with the exception of the conversion of iron oxide minerals into iron sulphide minerals.

7 Global Implications of a Hydrogeological Paradigm Shift

EWAN W. ANDERSON

Emeritus Professor of Geopolitics, University of Durham

Throughout time, human development has depended on water. In what remains a seminal work on the subject, *Man's Role in Changing the Face of the Earth* [Thomas, 1956] includes more references to water than to any other subject. In the *Voice out of the Whirlwind*, The Lord speaks to Job on man and his world [Job 38:1–41]. In an elegant poetical translation [Bates, 1936], the Voice asks . . .

> Who hath cleft a channel for the waterflood,
> or a way for the lightening of the thunder;
> to cause it to rain on a land where no man is;
> on the wilderness, wherein there is no man;

The Voice depicts the landscape, unencumbered by the works of man, in which the water follows the cleft. After further description of hydrological phenomena, the Voice states:

> The waters are hidden as with stone, . . .

This can be interpreted that the cleft is, at least in part, subterranean. Before the advent of drilling, its presence is therefore only likely to have been revealed by the occurrence of springs. Thus, the idea of groundwater and, in particular, groundwater flows in channels was predetermined almost from the beginning. Yet, this is the essence of the megawatershed and the linear fractured rock aquifers, which have raised the issue of a paradigm shift.

Modern Groundwater Exploration: Discovering New Water Resources in Consolidated Rocks Using Innovative Hydrogeologic Concepts, Exploration, Drilling, Aquifer Testing, and Management Methods, by Robert A. Bisson and Jay H. Lehr
ISBN 0-471-06460-2   Copyright © 2004 John Wiley & Sons, Inc.

Subterranean clefts can be translated in geological terms as faults, produced by movements within the Earth's crusts. This relationship between earthquakes and water flow was noted in the Classical World. In Meteorologica, Aristotle observed that:

> Water has sometimes burst out of the Earth when there has been an earthquake [Translation: Lee, 1952].

Similarly, Pliny in *Natural History* wrote:

> Earthquakes occur in a variety of ways, and cause remarkable conse-quences, in some places overthrowing walls, in others drawing them down into a gaping cleft, in others thrusting up masses of rock, in others sending out rivers [Translation: Rackham, 1938].

The relationship among earthquakes, clefts, and water springing from the ground, if not the processes involved, were thus recognized from the birth of science. Further examples, representing each stage of human development, can be seen in *Traces on the Rhodian Shore* [Glacken, 1967], generally accepted as the greatest geographical tome ever written. In it, Glacken follows the nature and culture in Western thought from ancient times to the end of the eighteenth century.

From the nineteenth century to the present, as more detailed knowl-edge has been accumulated and hydrology has been recognized as a separate subject, the predominant model that has developed has been that of the drainage basin or catchment with its attendant shallow groundwater storage, principally in alluvial deposits. Typically, the hydrological cycle, as one model, has been fitted to the drainage basin as a second model. The inflows have been assumed to result almost exclusively from precipitation, and the outflows are stream flow. The system as envisaged is therefore self-contained, and the basin is that defined by geomorphology at the Earth's surface. The aquifers in which groundwater is stored are distinguished as alluvial, those within the deposits of the river channel and permeable sedimentary rock with out-crops in the basin, which in turn lie above "impermeable" crystalline rocks.

At the same time, largely as a result of drilling for petroleum in the Middle East, the presence of deep "primary" aquifers has been estab-lished in thick units of sedimentary rocks based on the assumption of primary (intergranular) porosity (sandstones) and open cavities dis-solved from soluble layers within the sedimentary units (limestone) as the aquifer models. Using these models, seven major deep groundwater basins have been identified in the Middle East, which together under-lie almost 50% of the land area [Anderson, 2000]. It is conservatively estimated that 65,000 km^3 of stored groundwater exists in the MENA region. Points of contention introduced by the megawatershed model include not only the exact dimensions and boundaries of the sedimentary basins, but also solid evidence of intrabasin and interbasin secondary porosity and hydraulic interconnectivity resulting from regionally pervasive, deep-seated faults and fracture systems.

These observed megawatershed attributes introduce three major new elements into the traditional basin models: more water storage capacity, much higher permeabilities, and myriad underground channels that connect deep groundwater "primary" and "secondary" storage media directly with many types of active surface recharge. For example, big rivers (e.g., Nile) cross open fractures (cascades), and storm-fed wadis dispense water directly through cobbles into the deep-penetrating fractures that mostly define wadi geography (e.g., Oman, Somalia, Sudan), whereas open fracture systems in bedrock underlying thick, water-saturated alluvial deposits (e.g., Okavango Delta) receive potential recharge every time the hydraulic head increases. Obviously, the efficiency of these recharge mechanisms depends on many factors that must be addressed by competent investigators on a case-by-case basis. However, until the megawatershed model discovery, all of these vast deep aquifers were thought to contain fossil water with little or no access to recharge. Perspective really does matter a lot to the outcome of any endeavor in the natural sciences as with any other enterprise. In this case, the difference between traditional hydrological models and The Megawatershed Paradigm is simply that hundreds of cubic kilometers per year of renewable fresh water potentially exist under the feet of the driest places on Earth. Of particular relevance throughout this discussion is that the megawatershed model is not an academic concept, but rather an exploration paradigm. The principal underground pathways for groundwater catchment, storage, and migration in any given region can be identified and quantified using the megawatershed model and appropriate exploration technologies.

The traditional drainage basin concept has been shown to be inadequate, particularly in the context of its groundwater component. Indeed, in *Land, Water and Development*, a standard book on river basin systems and their sustainable management [Newson, 1992], a major conclusion is that:

The lessons are surely that hydrologists must start with policy in mind; just as engineering hydrology was adjusted to the service of society, so environmental hydrology must agree a certain subordination to human needs. It must anticipate needs. For example, at the time of writing it would seem obvious to put more effort into groundwater research since, during periods of environmental change, the groundwater store provides continuity and survival.

Although more knowledge of groundwater flows is required, there is virtually no detailed information on deep aquifers.

Thus, in a very real sense, the present work on megawatersheds provides a link between drainage basin theory and deep aquifer theory, with the great advantage that it is evidence-based research. On various scales, from those of the major region to the very local, field evidence has been assembled covering both the form and the process of the megawatersheds providing a detailed model from recharge to discharge.

PARADIGMS

In *The Structure of Scientific Revolutions*, Kuhn [1962] accords a key role in the evolution of science to the identification, development, and possible demise of paradigms. A paradigm sets the framework for science by indicating what is known, what is yet to be solved, and the procedures for solution. It therefore encapsulates a pattern of scientific activity that has relatively long-term stability.

In some ways, a paradigm is like an all-embracing model, but it differs from a model in that it is more wide ranging and less specifically formulated, and more important, it refers to patterns of searching the real world rather than to the real world itself [Haggett and Chorley, 1967]. If the approach to hydrogeology described in this book represents a paradigm shift, it is against these definitions that it must be tested. In this context, it must be realized that the megawatershed approach is more than a changed viewpoint on the hydrological cycle and the drainage basin in that its main focus is on the provision of greatly enhanced water supplies. Therefore, it will influence current paradigms in many other fields of human endeavor in what might be seen as a multiparadigm shift.

As is well known, most of the global water supply is in the ocean basins. Of that which is on land, a very high proportion is frozen. Using current models, it is reasonable to assess that 0.6% of global water occurs as groundwater and 0.02% in the form of rivers and lakes. Therefore, when precipitation occurs, the proportion entering the groundwater system is at least 30 times that which remains in channels and lake basins on the surface. Leaving aside the fact that the evidence from megawatersheds will require some enhancement of the groundwater proportion, the fact that groundwater, under appropriate conditions, is concentrated in large-scale fracture systems exhibiting complex, fractal architecture, completely alters ideas about water supply and all that depends on it. A useful analogy can be drawn with the qanat, falaj, or kareze of the Middle East, particularly Iran and Afghanistan. The qanat is a gently slopping subterranean tunnel excavated from a permanent source of water, the mother well in the nearest area of upland, to the village. It ensures a permanent supply of water for domestic and agricultural use. Megawatersheds-related deep bedrock aquifers are natural conduits on a much larger scale. The qanat can be compared with the well, which allows water to be drawn from the aquifer, in simple terms, an area of saturated permeable rock. If the well strikes a regional fracture zone, it is likely to have a higher yield and more rapid recovery rate after abstraction than a well that draws from locally recharged rock porosity or jointing, where in a dry period, the water table is likely to be lowered and wells run dry. Furthermore, local aquifer wells compete with one another, and all local farmers are probably drawing on the same shallow aquifer. In contrast, the qanat requires cooperation of a high degree. If one farmer fails to cooperate or abstracts more than his share of the water, all suffer. Thus, it is not unreasonable to consider that in the mists of past time, the decision to build a qanat represented a paradigm shift from the sinking of wells.

In terms of a hydrogeological paradigm shift, the issue must be raised as to how different is the megawatershed approach to the hydrological cycle/drainage basin concept. Viewed from one angle, this question includes elements of the catastrophism/uniformitarianism debate. A key to development of the megawatershed theory is the geological concept of tectonically induced, hydraulically conductive regional systems of fractures and faults in all types of consolidated materials, ranging from sandstones and limestones to basalts and metamorphic rocks.

In surface drainage basins, river valleys are the products of a long period of erosion, during which no variation in rock resistance, however minute, is considered insignificant. This is essentially true also of near-surface and deep faults and bedrock fracture zones that transmit and store water. Weathered crystalline rock in fault and fracture zones often produce thick, extensive primary aquifers that are hydraulically continuous with unweathered part secondary (fracture) aquifers that may persist for thousands of meters depth and hundreds of kilometers laterally. In short, both surface and subsurface water-bearing landscapes are molded by periodic catastrophic events and shaped by the incessant processes of weathering and erosion. This is the genesis of megawatersheds.

The interpretation of the hydrological cycle in the drainage basin varies. Traditional models describe outflow from groundwater storage into channel and subsequent stream flow. However, a number of investigators have observed evidence of deep underground outflow or leakage directly into offshore sediments on continental shelves [Moore, 1996] and slopes [Driscoll, 1997], as well as inland into low-lying playa lakes in heavily faulted basins [USGS Groundwater Atlas, Le Gal La Salle, 2001]. This remote submarine discharge evidently involves an equilibrium state of deep groundwater flux that requires continuous recharge to very large, deep groundwater systems over geological time to maintain a pressurized system. This type of regional "steady state" groundwater storage and flow concept strongly supports the mega-watershed model, as developed in current theory.

Furthermore, it is clear that in the drier parts of the world, river channels are likely to be effluent, with water moving in the channel well above the level of the water table from which it is separated by a phreatic zone. If the alluvial lining of the channel is incomplete or for some other reason ineffective, there will be downward leakage. In this context, to provide an idea of the scale of the leakage, it must be remembered that from the confluence with the Atbara River in Sudan to the Mediterranean Sea, there are no Nile tributaries and there is no currently known inflow. The average annual flow of the Nile is approximately 80 billion m^3 across an arid landscape. It is noted in *Water Resources in the Arid Realm* [Agnew and Anderson, 1992] that arid land groundwater is recharged naturally in three ways: (1) by exogenous rivers such as the Nile; (2) by seasonal flows from wadis and by springs; and (3) by runoff. Wadi flows and runoff are normally of very limited duration and are unlikely to account for major additions to the annual total. Springs may result from leakage. Thus, leakage from major perennial rivers is likely to constitute the main input to the groundwater system, as well as a new potential recharge mechanism, long-distance subterranean bedrock fracture flows within megawatershed domains that extend into high elevation rainfall catchments, which appears likely, from recent studies, to be the case in the northern range of the Nile.

Leakage is also taken into account in the water balance equation, which states that precipitation equals runoff plus evapotranspiration plus or minus changes in surface storage, soil moisture storage, aeration zone storage, groundwater storage, and deep transfer across the watershed. It is interesting that many of these variables are difficult to measure and the extrapolation from point measurements to the drainage basin as a whole is even more conjectural. Deep vertical and horizontal transfers cannot be measured directly using conventional means, and therefore, any currently available statistics given must merely represent the remainder after all other estimates have been taken into account. New methods of hydrological parameters measurement introduced in the context of exploring megawatershed domains are replacing conjecture with actual data for the first time in history, as the Trinidad and Tobago case studies illustrated in chapter 6 demonstrate.

It can be concluded that there is some recognition of leakage from the groundwater system within the traditional drainage basin model, and there are also inputs into deeper aquifers from effluent channels. If these leakages are taken into account together with the recognition that deep transfer across watersheds can take place, then there is already some indirect acknowledgment that there must be larger watersheds. However, the general consensus seems to remain that these are deep aquifers, the exact locations of which are relatively vague and the main storage of which is probably fossil water. Even if there is taken to be coincidence between megawatersheds and deep aquifers, the factors of recharge and discharge and the possible relationship to regionally extensive fault and fracture zones not only provide a completely different emphasis, but also offer a contrasting vision of water supply.

The situation is summarized in *The Middle East: Geography and Geopolitics* [Anderson, 2000].

Additional current research in the Levant has already begun bridging the conceptual gap between traditional models and The Megawatershed Paradigm by observing the fact that between the deep sources of presumably nonrenewable fossil groundwater and the surficial aquifers that are known to be replenished regularly occurs water in major faults such as the Jordan transform fault. This water has been shown by researchers to have been collected from fault zones in the mountains of Lebanon and Southern Turkey in what can be described as a major advance in recognizing geological controls of groundwater's rainfall catchment and recharge. The logical next step in the conceptualization process is to also recognize the common structural-hydraulic linkages (connectivity) between these major, deep-seated fault (and related open fracture) zones and the so-called "fossil" deep aquifers. In an active tectonic environment, all brittle rocks break in response to common stressfields, and groundwater will inevitably flow downgradient through major conductors into and through the very heart of the deep aquifers, creating a mélange of waters in a hybrid deep aquifer, combining new and ancient waters in close proximity, but not always freely mixing. The megawatersheds model explains most of the paradoxes of four decades of inexplicably diverse water ages and chemistries measured in the same deep aquifers.

The two-volume *Encyclopedia of Global Change* [Goudie, 2002] was produced by the world's scientists to celebrate the millennium. The *Encyclopedia* plots the relationship between environmental change and human society, and water in various guises is a major topic. In discussing drought, Agnew concludes: "There is a need to move away from responding to drought (tactical strategies) to a more proactive approach (strategic measures) for drought management and mitigation."

It is believed that the megawatershed approach is one (and arguably the most significant) such strategic measure, as further discussed by Alexander Raymond Love in Chapter 2.

ECONOMIC CONSIDERATIONS

It has already been established that, in terms of hydrogeology, there are grounds for considering that the megawatershed approach represents a paradigm shift. In fact, megawatersheds may also represent an economic paradigm shift. In this and the following sections, the effect of this approach is tested on other paradigms to ascertain whether the case for paradigm shift is enhanced or weakened.

Water is a fundamental resource and is crucial in the development of other resources, most obviously the range of agricultural products. It is estimated that over 1.2 billion people lack adequate clean drinking water and almost 3 billion lack access to adequate sanitation [Gleick, 2002]. The demand for water increases by between 3% and 4% each year, and apart from the issues of an increasing population and enhanced urbanization, other considerations include potential climatic change, the demands of the ecosystem, and the potential for conflict over water resources.

In its potable form, water is an economic factor in its own right, but it also exercises a major control over agriculture and is of significance in extractive and manufacturing industry. With regard to agriculture, its significance can be appreciated from the statistics of global food security. It is calculated that some 400 million people suffer from an acute lack of the necessary calories, protein, vitamins, and minerals needed to sustain their bodies and minds in a healthy state. More than 800 million people do not have enough food to meet reasonable nutritional needs. Furthermore, 14 million children under the age of five die each year from malnutrition, undernutrition, and related disease [Dando, 2002]. If these issues of water and food deficiency are to be effectively addressed, new approaches to water resources assessment, development, and management are required. If the quantity and quality of available water is to be significantly enhanced, a totally new approach to resources and resource identification is needed. From the evidence presented in this book, commencing with the Sudan economic development case study of 1989 to Trinidad in 2002, it can be seen that for the locations in which the megawatershed approach has been realized, there has been an order of magnitude increase in the supply of water and associated economic development opportunities in the cases where the model was used to discover sufficient new sustainable water resources to create fresh water self-sufficiency for an entire country. Furthermore, the quality of the water has not in general caused problems and the resulting damage to or stress upon the environment has been undetectable.

There is a close relationship between the availability of water and the pattern of agriculture. Rangeland pastoralism still requires water points for the animals, but the most intensive forms of agriculture tend to be closely associated with irrigation. Given sufficient water, it may be possible to obtain three crops in one year from the same piece of land. There is a clear positive correlation between the global map showing access to water and that depicting intensity of land use. It must be remembered that agriculture may be distinguished from other forms of production by a number of characteristics, including the biological nature of the production process. For this, water is crucial.

The classic model of the agricultural landscape is that of Von Thunen [Conkling and Yates, 1976]. In this, economic rent per acre was plotted against distance from the market in miles. The result was a series of concentric rings around the market with cultivation nearer to it and grazing further away. However, this model was posited upon the presence of an isotropic plane on which no environmental differences existed. Once water is introduced in the form of, for example, a river, both the agricultural and the settlement patterns change. Further, access to water adds value to the land, and thus, the pattern becomes considerably more complex. In the greater part of the developing world in which there is limited or scarce water, it can be seen that the introduction of a further source can result in a completely different agricultural landscape. If overall water availability is greatly enhanced, from multiple point-sources on a broad geographic scale as commonly happens from megawatersheds, then both the intensity and the type of agriculture are likely to undergo major alteration.

If agricultural production is to expand to alleviate food shortage, the irrigated area will require enhancement. Already, the area under irrigation has tripled in the past 50 years. By far the greatest area, 63% of the world-irrigated regions, is in Asia, whereas only 2% is in Sub-Saharan Africa. Barker and Seckler [2002] identify a new paradigm of water resource management that is gradually emerging to meet the twin challenges of growing water scarcity and competition for water. The first component of the new paradigm is an increased recognition of the need to recycle and reuse water in river basins, and the second component focuses on recent advances in information technologies. A third component can now be added: to strategically develop mega-watersheds on a regional scale to maximize economic and public health benefits. For the first time in history, farms and settlements can thrive in areas independent of perennial rivers, springs, and whimsical local climates.

The rapid development of techniques such as remote sensing and geographic information systems will enable water managers to deter-mine water needs, and the megawatersheds model will assess avail-ability both in the surface and subsurface areas, at a scale transcending the river basin. This reference encapsulates the procedure by which the megawatersheds approach has been developed. To facilitate the iden-tification of appropriate locations, both remote sensing and geographic information systems (GIS) are basic requirements.

In the context of irrigation, a warning note has been sounded about salination problems. As the area under irrigation is extended and a wide range of different techniques for water abstraction and application are employed, salt levels can build in the soil causing salinization [Goudie, 2002B]. The problem can be addressed through eradication, conserva-tion, or control. The main approach to eradication is through the appli-cation of large volumes of low-TDS water to leach out the salts to an acceptable level. In certain specific areas, especially in the Middle East, water from deep aquifers in megawatersheds might offer a viable pro-cedure for doing this.

For agriculture, the megawatersheds model offers a number of advantages. Like the qanat, the system is essentially cooperative rather than competitive, and with known flows, costs can be shared more equi-tably over a larger area. As flow is unaffected by seasonality, there is far greater control over water use and therefore over the crops that can be grown. Agricultural practices may involve low technology or high technology, and the costs of speculative drilling would be eliminated. In fact, the use of megawatersheds models to develop strategic water sources fits perfectly with the economic development concept of optimum utilization of existing infrastructure. With the megawatersheds model, new production wells can be sited to take advantage of existing water delivery systems, commonly built in the most heavily populated and farmed areas, so that water delivery can commence within months instead of years, and at a much-reduced cost. In many areas of the Middle East where qanat-type structures exist, they will become major water conveyors once again, as Alexander Love describes elsewhere in this book, combining ancient and space-age technologies to the greatest benefit at the least cost.

For extractive industry in which there is large-scale water use such as for cooling, pumping, or washing, additional water supplies have an obvious advantage. Location of a plant in relation to to megawatershed supplies with a large and regular water discharge allows all procedures to occur on site. Thus, the costs involved in bringing water in or in transporting untreated raw materials away are removed. These may be special cases in that the location of such a plant must be fixed by the quality and quantity of the raw material available, but the manufacturing industry can generally locate according to the economic requirements. Locational decisions in the manufacturing industry can be contrasted with those in agriculture. In the former, the product is known and the location can be varied, whereas in the latter, the location is normally fixed and the product can be varied. Therefore, for the manufacturing industry, water can be a locational factor, whereas for agriculture, it is more likely to influence the type of land use and the specific products. Clearly, sources of water affect settlement patterns and density and these affect the manufacturing industry in that location is likely to be somewhere between the raw materials and the market. Depending on the other factors of production such as raw material, power, and skilled labor, the optimum location, facilitated by additional water supplies, would be in the market. An extreme case where this is obtained is that of certain well-known drinks that travel the world as powders and are reconstituted in bottles as near as possible to the market. An example of a megawatershed application in this context is a 1988–1990 USAID-funded project that coupled The Megawatershed Paradigm with modern economic development analysis methods to evaluate untapped groundwater resources potential together with actual water-constrained mining, agriculture, and industrial development businesses located in the Red Sea Province of the Sudan. Groundwater maps, a technical report, and business plans for gold, cement, and marble extractive industries, agriculture, and urban development were generated by co-contractors BCI-Geonetics, Inc. and Deloitte Haskins & Sells under a cooperative agreement with the Khartoum-based Sudanese Company for Water & Economic Development and USAID.

The basic and still best known model of the industrial landscape is that of Weber [1929]. This is a least-cost model constructed to indicate industrial location under certain specific conditions, among which is the fact that fuels and raw materials that require production are fixed in location and not ubiquitous. Therefore, the addition of water or the introduction of a new source has a significant effect on industrial location theory. Furthermore, Weber introduced other conditions such as the loss of weight during manufacture and distribution costs compared with transport costs. Clearly, water is a heavy and bulky material to transport, and therefore, if water is any way dominant in the process, the location will be raw material orientated. In general, the influence of water on the location of extractive and manufacturing industry in remote areas (Sudan USAID model project) will be as economically as important as upon agriculture. Overall, the megawatersheds approach is of great significance in the economic landscape, especially in developing countries.

SOCIAL CONSIDERATIONS

The availability of potable water is, under normal circumstances, the prime consideration for settlement. The wetpoint site is predominant virtually everywhere except in marshy areas where the drypoint site becomes important. Throughout the temperate world, villages have been constructed along spring lines and watercourses. In arid and semi-arid environments, the settlement pattern relates to wells, oases, river valleys, and wadi courses. Particularly in the developing world, it is obvious that the identification of regional deep groundwater systems will allow production of a new reliable source of water, giving rise to new related settlement patterns or the reorientation of preexisting patterns. Furthermore, the occurrence of temporary and migratory patterns is declining and a new source would bring stability if not necessarily a sedentary way of life. GIS-based megawatershed maps will encourage and actually enable planners to "think globally and act locally."

If the new source was associated with megawatersheds, the need to endure seasonal fluctuations from dry periods to floods would be ended. Indeed, in the extreme case, the need for settlement along rivers that flood regularly and disastrously could be greatly reduced or eliminated.

It must be remembered that the urban revolution occurred in areas such as Mesopotamia in which there was some control over water supplies. In the case of megawatershed-source water, there would be a great deal more control. Also, although there might in some cases be a higher incidence of dissolved minerals, the water would be pollution free.

In can be seen that with the large-scale development of megawatersheds, the control of rivers, springs, and wells over settlement patterns would be at least lessened and in some cases removed. The other obvious influence of water on settlement is through the economic landscape. This is seen classically in the model of Christaller [1933, translation 1966] modified by Losch [1943, translation 1954]. The basis for the pattern of villages and towns was a central place or market around which a uniform pattern of lesser settlements developed. As with Weber, the fundamental platform was an isotropic plane. Therefore, the introduction of a new water supply, for example, a series of wetpoints in a megawatershed would completely distort the model, because there would be competing access not only for urban services, but also for water supplies. Furthermore, owing to the agricultural differences resulting from the additional water, there would be a marked differentiation in services provided by different central places.

Clearly, in the social context, lifestyles and locations for settlement are highly dependent on available water and would therefore be significantly changed should a new source be introduced. Thus, the possibility for change in the social landscape, should the full potential of megawatershed-sourced water be realized, is enormous.

SECURITY CONSIDERATIONS

In that it is vital for life and economic development, water must be considered a strategic resource. Indeed, it is more fundamental than petroleum and various types of minerals that are normally considered strategic. Like other economic minerals, water's end use is critical and it is designated strategic when there are limitations on supply. As with other strategic minerals, water can be recycled and there is huge potential for conservation measures. However, unlike most strategic minerals, there is no substitute. Desalination might be considered a form of substitute in coastal and some inland areas, but it is expensive and, as an option, is only open to certain states. This can be illustrated by the fact that some two-thirds of its installed capacity is located on the Arabian Peninsula.

As it is so vital to most facets of human life, water security is a key issue for all societies. Groundwater from deep sources is more secure than surface water in that diversion of flow is not practical and over-abstraction has historically been almost impossible to gauge (although debates on the subject rage continuously). Turkey has been accused by Syria and Iraq of overabstraction from the River Euphrates as a result of the South-East Anatolian Project and in particular the Ataturk Dam. Libya, Egypt, and Sudan share the Kufra or Western Desert Deep Aquifer, but Libya has not yet been (officially) accused of over-abstraction. However, a new international initiative is now underway to map and understand the great aquifer systems of North Africa [CEDARE, 2001], and the biases of the models employed during that exercise will in some part determine whether groundwater will remain politically serene. Certainly, groundwater obtained from a data-driven megawatershed model introduces new predictability and enhances stability in a historically data-less and self-interest-driven debate and lends credence to the concept that knowledgeable use of deep ground-water sources will provide much more security than any surface source can offer.

Properly managed water production from deep wells in megawater-sheds is sustainable, and water can be withdrawn at a constant, pre-dictable rate without seasonal fluctuations, because groundwater withdrawals, storage, and recharge are maintained in balance. This allows not only security, but also long-term planning, and if the analogy of the qanat is taken, it demands international cooperation. The pro-duction wells are drilled and sealed into deep groundwater water sources of natural high quality, and so they are resistant to contamina-tion and sabotage by terrorists. Most wells drilled into megawatersheds are high yield and have sufficient economic value to justify high-security capital works construction methods and close monitoring. Deep aquifer access is limited to the wells themselves, and limiting and safeguarding the number of access points can avoid deliberate damage to aquifers. The megawatershed model is GIS-based and forms a plan-ning tool for this type of advanced planning to optimize water security.

In broader terms, deep groundwater systems contain thousands of cubic kilometers of water in storage and subterranean water flows within the vast domains, and from myriad areas of recharge are extremely slow compared with river flows, even in the secondary (frac-ture) hydraulic networks of the megawatersheds. Therefore, there should be no crisis caused by perceived near-term competition for the resource among states, and there would be ample time for knowledge-based fair and peaceful hydropolitics to emerge. Hydrology and politics are combined in the term hydropolitics and together illustrate the codependency of the two topics and subtlety of the subject. In practice, both groundwater hydrology and politics have been more art than science (some would say black arts), both based very largely on perceptions. Hydrogeology is a discipline within the science of hydrology that deals with perceptions of deep-earth realities that have been largely unmeasurable and therefore "assumed" until the advent of modern geological concepts and recent technological advances in groundwater exploration that led to the megawatershed concept. In negotiating over water rights, politicians can be relatively confident from hydrologists' reports about the average flow of a river and know that the flow varies over very short periods of time, so the term "average flow" may be meaningless, which can lead to hot debates about damming or diverting rivers based on average flow statistics. In the case of political discussions of deep groundwater flows and extractions, the political perception of groundwater flows is based on hydrogeologists' preconceived perceptions of a poorly understood, mostly unexplored, and virtually untested environment. If the hydropolitics of groundwater continue to heat up in coming years and groundwater hydrologists' perceptions of deep aquifers continue as before, the current, inherently unstable hydropolitical milieu will grow worse and potentially disastrous decisions may be made about the future fate of water use in the MENA countries and elsewhere.

LEGAL CONSIDERATIONS

Given the significance of water security, water rights and legal access to water become an important issue. International rivers are defined as those that occupy drainage basins shared by two or more states or that constitute the boundary between the states. In the case of the River Euphrates, it rises in Turkey and flows into Syria and then to its mouth in the Persian-Arabian Gulf through Iraq. The Euphrates is therefore known as a successive river. The River Tigris provides at one place the international boundary between Syria and Turkey and, immediately downstream, that between Syria and Iraq. It is therefore known as a contiguous river. In both cases, more than one state has claim on the water. Water law has developed in terms of surface flow but has produced a number of alternatives rather than a definitive internationally agreed position. For example, the principle of absolute territorial sovereignty, according to which a state has the right to use the river water that lies within its territory without any limitation, clearly conflicts with the principle of absolute territorial integrity by which no state may use the waters of an international river in a way that could cause detrimental effects to a coriparian. As a result of these opposing stances, there have developed the intermediate theories of common jurisdiction by all riparians over the entire drainage basin and the principle of equitable utilization, which permits the use of water in a way that does not harm other riparians. This last principle is expressed in the Helsinki Rules constructed by the International Law Association. The Helsinki Rules have now become the accepted legal foundation for the utilization of international rivers, and a full discussion on these and all other aspects of the subject is set out in Kliot [1994].

Rivers, whatever the difficulties of flow measurement, require the intervention of international law to settle disputes over abstraction. The effect of this is that water is seen as an object for possible conflict. In very few cases has water been envisaged as the focus for collaboration. One such collaboration resulted in the Indus Waters Treaty [1960]. Negotiations for this and the importance of the World Bank as a mediator are set out detail by Alam [2002]. However, the point remains that surface water that flows between countries can produce dispute over its utilization.

For subsurface water, there is increasing need for some sort of water law, but its imposition would raise exceptional complexities. In megawatershed environments with very large storage and extensive active recharge mechanisms, given the rates of transmissivity through deep aquifer rocks, it would be extraordinarily difficult to prove that abstraction from one point affected availability at another, at least in the short term. However, at the Oman/United Arab Emirates (UAE) boundary near Buraimi, the water table was in the 1980s declining at a rate of several meters a year. Recharge was from the mountains of Oman, and the bulk of the abstraction was for agri-business near Al Ain in the UAE. This circumstantial evidence of the case appeared relatively clear cut but still could not be proved by scientific evidence using traditional models and methods. Like shallow aquifers, deep aquifers frequently extend across international boundaries, but unlike shallow aquifers, their geology remains somewhat conjectural throughout most of the world. At depths of 1000 m and more, the transmissivity of rocks can vary widely and is largely unknown and ignored in traditional hydrological models because most data from myriad oil well logs, deep mine shafts, and deep wells drilled in many government-investigated areas like the Nevada atomic test site have not been applied in context with deep groundwater exploration or modeling until the advent of The Megawatershed Paradigm, which drew heavily from such previously unexploited databases. Also, in many instances, there may or may not be actual hydrological connections between various parts of what are now classified as "regional" aquifers like the famous Ogallala in the Western United State. These issues and in particular the question of law and groundwater are identified by Anderson [1995]. In the context of the current volume, Anderson continues:

> In the case of the megawatershed paradigm, the hydrology of deep groundwater systems may well be accurately ascertained, and the disposition of fractured rock aquifers or known pathways for groundwater recharge and migration will effectively compel cooperation between neighboring states and therefore legal agreements are necessary. With the megawatersheds approach, complex fracture network flow concentrations can be broadly identified in a fashion that bears some metaphorical resemblance to underground "rivers"; international law could be applied using the same criteria already established for surface flow. However, since the hydrology and abstraction are considerably more complex than with conventional rivers, there would need to be prior agreement to perform exploration in order to establish objective data for cooperation.

It can be seen that from the legal standpoint that megawatershed-generated flow presents many advantages for the clear utilization of the law over aquifers. Furthermore, unlike rivers, concentrated subsurface flow would, as a result of the technology requirements, demand collaboration of a high order, but would also permit adequate time to accomplish scientific goals.

MILITARY CONSIDERATIONS

An extension of security concerns and closely related to legal and political considerations is the military dimension. As Mark Twain observed:

> Whiskey is for drinkin'; water is for fighting.

Water is critical and where it is located also strategic. Because of scarcity, issues of access and denial develop. Once water enters the field of realpolitik, more commonly known as hydropolitics, tensions can build that may erupt into conflict. There have been several books about water wars, but throughout history, there have been very few battles that could be attributed solely to water. Gleick [1995] provides a useful summary of such events. In modern times, dams for the generation of hydroelectric power have been bombed in World War II and the dams of the Yalu River were bombed during the Korean War. In the late 1960s, the irrigation systems of North Vietnam were attacked by the United States. Very recently, a sizeable portion of the desalination capacity of Kuwait was destroyed by Iraq during the Gulf War of 1991. These are all examples of denial.

There have also been concepts for using water aggressively by releasing large quantities and damaging downstream structures. A good example, combining fact and perception, occurred in the Korean Peninsula before the Seoul Olympic Games of 1988. The northern branch of the Han River rises in North Korea, and in 1986, plans were announced to build the Kumgansan Dam on it. This raised the specter of a water bomb hanging over Seoul at a time when it would be most exposed to world publicity. Release of water from the dam, it was estimated, would cause a rise in river level of some 50 m in the city of Seoul, as a result of which the disruption of the rowing events would be a relatively minor consideration. South Korea took evasive action by building a series of levees and check dams. In time, imagery indicated that the construction in North Korea was far smaller than had been anticipated and indeed that perceptions had rapidly overtaken scientific facts.

However, the greatest potential for conflict over water flow has always been thought to concern the three major basins of the Middle East: the Nile, the Tigris-Euphrates, and the Jordan. In 1974, Iraq threatened to bomb the al-Thawra Dam in Syria and massed troops along the border. This was the nearest that the Tigris-Euphrates basin coriparians have come to actual conflict. With the construction of the Ataturk Dam (1990), the flow of the Euphrates was interrupted for one month so that the interstices of the dam could set. Additional water had been sent down river in advance so that overall there would be no loss to the downstream states. Nonetheless, the act of switching off flow was a highly significant hydropolitical event. The explanations of hydraulics' engineers and hydrologists from all three countries concerned were disregarded. The political leadership of Syria and Iraq met, despite frequently expressed mutual hostility, and protested that Turkey had ultimate control in that it could deny water or flood the downstream cities as it wished. In fact, the Mesopotamian Plain is far from ideal for using water as a weapon in that it is largely flat and denial of water has a definite time limitation in that storage behind the Ataturk Dam is limited. As the dam is also used for the generation of hydroelectric power, the lake needs to be maintained at a relatively high level, leaving little capacity for additional storage.

These examples illustrate some of the more obvious military considerations with regard to water. They all concern surface water because knowledge of groundwater among hydrologists let alone politicians is relatively limited. Obviously, there have been examples of well poisoning, but the effect of these is limited both in extent and time and can be mostly avoided using modern techniques. Water from megawatersheds combines many of the advantages and few of the disadvantages of that found in rivers or shallow aquifers. Access would require relatively high technology, and access points could be protected. At a time when a war on something as abstract as terrorism can be announced, there can be no guarantees that a war over a megawatershed would not take place, but it seems at best highly improbable. Therefore, it is reasonable to conclude that should the megawatershed approach prevail, water wars would be precluded.

As Anderson [2002] concludes in the context of conflict:

> To date, the most effective form of persuasion has been financial. Since major water projects are well beyond the means of most countries, international funding is necessary, and it is most frequently sought from the World Bank, which has been able to develop a policy of refusing support when there are clearly conflicting viewpoints about a project.

POLITICAL CONSIDERATIONS

As the political is the overriding decision-making system, political considerations include most of the arguments already deployed. In particular, the prime duty of a state is to safeguard the security of its inhabitants. The state should also be the source of law and in control of the military. Increasingly, politics and economics have become interdigitated, and it has already been demonstrated that social and economic concerns have a considerable overlap. Politics is about decision-making, and geopolitics takes into account the geographical elements that can influence decisions. Water as a crucial socioeconomic variable is geopolitically highly significant. For example, in the Middle East Peace Process, is it possible to see an accord between Israel and Palestine that does not take water into account? More likely, issues of groundwater supplies from the West Bank inhibit the whole idea of "land for peace."

In essence, a state comprises territory and an organized population. The relationship between water and population distribution has already been considered. With regard to territory, it needs to be defined by a boundary. Without a boundary, there can be no control, and without control, it is doubtful whether sovereignty in the fullest sense can be enjoyed.

Water is of particular relevance in boundary settlement in that there are obvious advantages in having an entire catchment within the territory of one state. Therefore, watershed boundaries are relatively common. However, rivers also provide boundaries in that they are linear features that are particularly obvious on the landscape. There are many disadvantages in that rivers are sources of water, possibly power, and transport, and the people living on either side are likely to have more in common than that which separates them. Therefore, a drainage basin is a natural unit both physically and in terms of human geography. However, psychologically rivers form an excellent boundary in that they indicate very clearly on the ground the extent of territory. Furthermore, crossing them violates territory. Therefore, if there is any doubt as to troop movements in the boundary zone, once a river is crossed, all doubts are erased. It is no surprise therefore that when boundary disputes are negotiated, physical geography and, in particular, hydrology plays a notable part.

There is no comparable political significance in megawatersheds as the source of water. The fractal architecture of natural bedrock fracture aquifers in regional tectonic frameworks subsurface drainage presents few clear surface expressions of its geography, save occasional geomorphically prominent bounding faults that may divert groundwater upward or laterally, but frequently they possess little groundwater development potential, and no faultline has deliberately been proposed as an international land boundary. In the case of maritime boundaries, it is true that geophysics plays a greater role in what is otherwise a featureless landscape. Politically, therefore, megawatershed-source water is only likely to enter the arena for decision-making with regard to cooperation over abstraction. The prospects of the immense water supply potential of the Levant's huge fault and fracture-system-derived megawatersheds influencing the political disputes among Israel, Syria, Jordan, and Palestine are obviously high.

As water is such a crucial component of territory and territory is the fundamental platform of the state, the introduction of a relatively large, sustainable, nonfluctuating supply can alter the political balance.

CONCLUSION AND PROSPECT

If the megawatershed approach to hydrological and hydraulic problems can be successfully prosecuted on a wide scale, there seems little doubt that it will represent a hydrogeological paradigm shift. More than that, it will have the capacity to alter most of the basic models that have been designed to describe human occupancy of the landscape. It will also influence security and the attendant military and legal issues. At this stage, it cannot be argued that for each consideration there would be a paradigm shift, but for some that would be the case and for others at least the basic model would be amended. The potential is clearly enormous. It is hoped, therefore, that the procedure can achieve universal acceptance so that a global regional analysis can be mounted to identify megawatersheds, particularly in areas of water stress. A database and a GIS with worldwide coverage would add significantly to the effectiveness of agencies concerned with conflict resolution, disaster relief, and the targeting of aid.

BIOGRAPHY

Ewan W. Anderson is Emeritus Professor of Geopolitics at the University of Durham and Visiting Professor at the Institute of Arab and Islamic Studies, University of Exeter. He also has appointments at the Universities of York (UK), Georgia (US), and Al Akhawayn (Morocco). Professor Anderson's career interests focus on the economic security and social well-being of nations and people, a mission supported by undergraduate and advanced degrees in geography, politics, geomorphology, hydrology, and childcare.

Professor Anderson has worked on water issues for the governments of Oman, Jordan, and Turkey. He directed development projects in several countries in Africa and the Middle East including one for the UK Department of International Development on targeting aid in Northern Iraq (1996–2001).

From 1988 until 1996, Professor Anderson was strategic analyst to the Supreme Allied Commander Europe. In 1996, he was appointed as a political specialist to the British military Higher Command and Staff Course. He was the chief *rapporteur* and analyst for the series of NATO conferences on geopolitics held at the Marshall Centre, Garmish. Since 1982, Professor Anderson has worked on seven international boundary settlements and was from 1994 to 2000 in charge of non-legal boundary research for the government of Saudi Arabia. Currently, he is working with the Ministry of Defence (UK) on boundaries and global terrorism.

He has published fourteen books and some 250 papers in his chosen disciplines, including many on international strategic minerals and water-related issues ranging from the geopolitics of water and water security to groundwater hydrology.

REFERENCES

Alam, U.Z., 2002, Questioning the Water Wars Rationale: A Case Study of the Indus Waters Treaty. The Geographical Journal, Vol. 168, No.4, pp. 341–353.

Agnew, C.T., 2002, Drought: An Overview, in Goudie, A. S. (Ed.) Encyclopedia of Global Change: Environmental Change in Human Society, Vol. I, Oxford University Press, New York, pp. 274–278.

Agnew, C.T., and Anderson, E.W., 1992, Water Resources in the Arid Realm, Routledge, London.

Anderson, E.W., 1995, Water Resources in the Middle East: Boundaries and Potential Legal Problems, Using Jordan as a Case Study, in Blake, G.H., Hildesley, W.J., Pratt, M.A., Ridley, R.I., and Schofield, C.H. (Eds.): The Peaceful Management of Transboundary Resources, Graham and Trotman/Martinus Nijhoff, London, pp. 203–208.

Anderson, E.W., 2000, The Middle East: Geography and Geopolitics, Routledge, London.

Anderson, E.W., 2002, Middle East, in Goudie, A.S. (Ed.): Encyclopedia of Global Change: Environmental Change in Human Society, Vol. II, Oxford University Press, London Youk, pp. 87–92.

Aristotle, 1952, Meteorologica Translated: Lee, H.D.P.: Loeb Classical Library, Harvard University Press, Cambridge, MA.

Barker, R., and Seckler, D., 2002, Irrigation, in Goudie, A.S. (Ed.): Encyclopedia of Global Change: Environmental Change in Human Society, Vol. I, Oxford University Press, New York, pp. 706–710.

Bates, E.S., (Ed.), 1936, The Bible Designed to be Read as Living Literature, Simon and Schuster, New York.

CEDARE, 2001, Nubian Sandstone Aquifer System Programme: The Centre for Environment and Development for the Arab Region and Europe.

Conkling, E.C., and Yates, M., 1976, Man's Economic Environment, McGraw-Hill, New York.

Christaller, W., 1966, Central Places in Southern Germany, Translated: Baskin, C.W., Prentice-Hall, Englewood Cliffs, NJ.

Dando, W.A., 2002, Food, in Goudie, A.S. (Ed.): Encyclopedia of Global Change: Environmental Change in Human Society, Vol. II, Oxford University Press, New York, pp. 455–462.

Glacken, C.J., 1967, Traces on the Rhodian Shore, University of California Press, Berkeley, CA.

Gleick, P.H., 1995, Water and Conflict, in Lynn-Jones, S.M., and Miller, S.E. (Eds.): Global Dangers: Changing Dimensions of International Security, MIT Press, Cambridge, MA.

Gleick, P.H., 2002, Water, in Goudie, A.S. (Ed.): Encyclopedia of Global Change: Environmental Change in Human Society, Vol. II, Oxford University Press, New York, pp. 494–502.

Goudie, A.S. (Ed.), 2002A, Encyclopedia of Global Change: Environmental Change in Human Society, Vol. I and II, Oxford University Press, New York.

Goudie, A.S., 2002B, Salinisation, in Goudie, A.S. (Ed.): Encyclopedia of Global Change: Environmental Change in Human Society, Vol. II, Oxford University Press, New York, pp. 341–346.

Haggett, P., and Chorley, R.J., 1967, Models, Paradigms and the New Geography, in Chorley, R.J., and Haggett, P. (Eds.): Models in Geography, Methuen, London.

Kliot, N., 1994, Water Resources and Conflict in the Middle East, Routledge, London.

Kuhn, T.S., 1962, The Structure of Scientific Revolutions, University of Chicago Press, Chicago, IL.

Losch, A., 1954, The Economics of Location, Translation: Woglom, W.H., Yale University Press, New Haven, CT.

Moore, W.S., 1996, Large Groundwater Inputs to Coastal Waters Revealed by ^{226}Ra Enrichment: Nature, Vol. 380, pp. 612–614.

Newson, M., 1992, Land, Water and Development, Routledge, London.

Pliny, 1938, Natural History, Translated: Rackham, H., Loeb Classical Library, Harvard University Press, Cambridge, MA.

Thomas, W.L. (Ed.), 1956, Man's Role in Changing the Face of the Earth, Vol. I and II, The University of Chicago Press, Chicago, IL.

Weber, A., 1929, Theory of the Location of Industries, University of Chicago Press, Chicago, IL.

Index

*Modern Groundwater Exploration: Discovering New Water Resources in Consolidated
Rocks Using Innovative Hydrogeologic Concepts, Exploration, Drilling, Aquifer
Testing, and Management Methods*, by Robert A. Bisson and Jay H. Lehr
ISBN 0-471-06460-2   Copyright © 2004 John Wiley & Sons, Inc.